Building Strategies
for GED Success

Science

Steck Vaughn™

A Harcourt Achieve Imprint

www.Steck-Vaughn.com
1-800-531-5015

STAFF CREDITS

Design: Amy Braden, Deborah Diver, Joyce Spicer
Editorial: Gabrielle Field, Heera Kang, Ellen Northcutt

PHOTOGRAPHY

Page 24r ©Biophoto Assoc./Photo Researchers; p.24l ©Roger Harris/Photo Researchers; p.42 ©Gianni Dagli Orti/ CORBIS; p.44 ©Frans Lanting/CORBIS; p.50 ©Susan Ruggles/Index Stock; p.52 ©John Coletti/Index Stock; p.61 ©Walter Stuart/Index Stock; p.74 ©CORBIS; p.131 ©Edward Kinsman/Photo Researchers.

Additional photography by Photos.com; ©Royalty-Free/CORBIS.

ILLUSTRATION

Franklin Ayers pp.11, 60, 62, 68, 76, 78, 80, 81, 82, 84, 85b, 89, 91, 106, 110, 111, 115, 116, 118, 123, 124, 126, 127, 128, 130, 132, 136, 143; Jonathan Massie pp.33l, 55; Rich Stergulz pp.46, 51; Wilkinson Studios pp.4, 143. All other art created by Element, LLC.

ISBN 1-4190-0799-8

Contents

To the Learner

Congratulations! You have taken an important step as a lifelong learner. You have made the important decision to improve your science skills. Read below to find out how Steck-Vaughn *Building Strategies for GED Success: Science* will help you do just that.

- Take the **Pretest** on pages 3–9. Find out which skills you already know and which ones you need to practice. Mark them on the **Skills Preview Chart** on page 10.

- Study the four units in the book. Learn about life science, earth and space science, chemistry, and physics. Check out the **GED Tips**— they've got lots of helpful information.

- Complete the **GED Skill Reviews** and **GED Strategy Reviews**. You'll learn a lot of important reading, thinking, and test-taking skills.

- As you work through the book, use the **Answers and Explanations** at the back of the book to check your own work. Study the explanations to have a greater understanding of the concepts. You can also use the **Glossary** on pages 152–157 when you want to check the meaning of a word.

- Review what you've learned by taking the **Posttest** on pages 143–150. Check the **Skills Review Chart** on page 151 to see the progress you've made!

Setting Goals

A goal is something you want to achieve. It's important to set goals in life to help you get what you want. It's also important to set goals for your learning. So think carefully about what your goal is. Setting clear goals is an important part of your success. Choose your goal from those listed below. If you don't see your goal, write on the line. You may have more than one goal.

My science goal is to

- get my GED
- improve my job skills
- get a new job that uses science

_____

A goal can take a long time to complete. To make achieving your goal easier, you can break your goal into small steps. By focusing on one step at a time, you are able to move closer and closer to achieving your goal.

Steps to your goal can include

- understanding science vocabulary
- reading science articles in the newspaper
- reading science magazines
- understanding science programs on TV
- helping your children with their science homework

We hope that what you learn in this book will help you reach all of your goals.

Now take the _Reading Pretest_ on pages 3–9. This will help you know what skills you need to improve so you can reach your goals.

Science Pretest

This pretest will give you an idea of the kind of work that you will be doing in this book. It will also help you to figure out which science skills you need to improve.

You will read short science passages and graphics such as graphs, maps, and diagrams. You will also answer multiple-choice questions. There is no time limit for this test.

Read each selection and question carefully. Circle the number of the correct answer.

Questions 1–3 are based on the paragraph and diagram.

The body is organized from smaller to larger parts. Cells are the smallest body parts. Cells working together are called tissues. Tissues form organs such as the heart and lungs. Organs carry out body functions.

1. Which are the smallest parts of the body?

    (1) cells
    (2) organs
    (3) tissues
    (4) organ systems

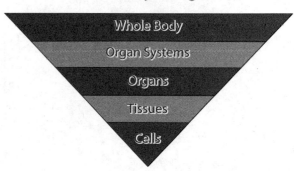

How the Body is Organized

2. According to the diagram, organ systems work together to support the

    (1) cells.
    (2) tissues.
    (3) organs.
    (4) whole body.

3. What is the main idea of the paragraph?

    (1) The heart and lungs are organs.
    (2) Cells are the smallest body parts.
    (3) Organs work together to carry out body functions.
    (4) The body is organized from smaller to larger parts.

Questions 4–6 are based on the paragraph and diagram.

There are three kinds of blood vessels: arteries, veins, and capillaries. The arteries carry blood away from the heart, and the veins carry blood back to the heart. The capillaries deliver the blood to the body's cells. The blood is always circulating in the body. For this reason, the heart and the blood vessels are known as the circulatory system.

Types of Blood Vessels

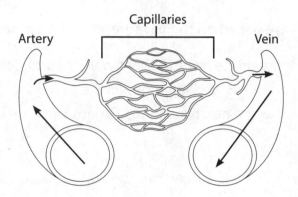

Arrows show the path of blood flow.

4. The heart and blood vessels are known as the

    (1) blood system.
    (2) circulatory system.
    (3) respiratory system.
    (4) capillary system.

5. According to the diagram, blood moves from the

    (1) veins to the arteries.
    (2) capillaries to the arteries.
    (3) arteries through the capillaries to the veins.
    (4) veins through the capillaries to the arteries.

6. What is the main idea of the paragraph?

    (1) Capillaries deliver blood to cells.
    (2) There are three kinds of blood vessels.
    (3) Blood is always circulating in the body.
    (4) Arteries carry blood away from the heart.

Questions 7–9 are based on the following map.

Desert Areas of the World

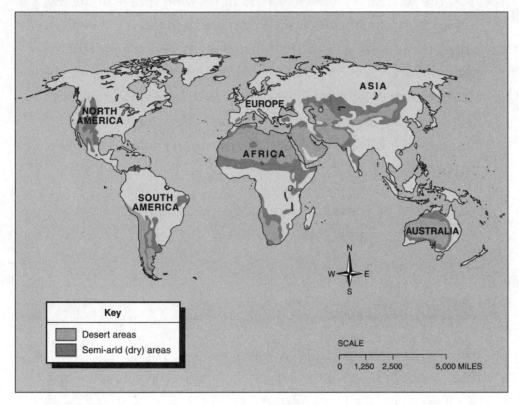

7. Which continent is almost all desert or semi-arid?

    (1) North America
    (2) South America
    (3) Africa
    (4) Australia

8. Which continent has no deserts?

    (1) Asia
    (2) Africa
    (3) Europe
    (4) South America

9. Where is the largest desert area located?

    (1) Australia
    (2) North America
    (3) South America
    (4) Africa

Questions 10–12 are based on the following paragraph.

The sun usually looks like a round, glowing sphere in the sky. Although the sun may look small from Earth, in fact it is the largest object in our solar system. The sun makes up 99 percent of the solar system. It also gives off an immense amount of light and heat. This large amount of light and heat give us warm, sunny days on Earth. The light and heat has another important effect. It is what makes it possible for living things—such as plants and animals—to exist on Earth.

10. Based on the information in the paragraph, you can conclude that the word *sphere* means

    (1) hot.

    (2) large.

    (3) ball shaped.

    (4) star shaped.

11. What percent of the solar system does the sun make up?

    (1) 1

    (2) 25

    (3) 50

    (4) 99

12. Based on the information in the paragraph, the word *immense* probably means

    (1) small.

    (2) large.

    (3) unusual.

    (4) unimportant.

Questions 13–14 are based on the following paragraph and bar graph.

Harmful chemicals in the air are called air pollution. Air pollution has both natural and human causes. For example, volcanoes put chemicals called sulfur oxides in the air. Sulfur oxides cause acid rain and harm plants and animals. However, volcanoes are not the main cause of acid rain. Humans are the main cause of acid rain. Human activities, such as burning coal in power plants and paper mills, put ten times more sulfur oxides in the air than natural activities.

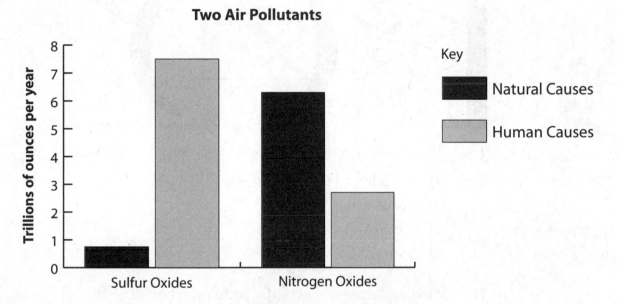

Two Air Pollutants

13. Which of the following is a main cause of acid rain?

    (1) natural activities
    (2) harmful plants
    (3) burning coal
    (4) erupting volcanoes

14. According to the bar graph, how much nitrogen oxide is released in the air by natural causes each year?

    (1) less than 1 trillion ounces
    (2) 2.7 trillion ounces
    (3) 6.3 trillion ounces
    (4) 7.5 trillion ounces

Questions 15–17 are based on the following paragraph and diagram.

A flat mirror reflects light at the same angle at which the light strikes it. The reflected image appears to be the same size as the object. A concave mirror is curved inward like the inside of a bowl. Light is reflected and then spreads out. The image appears larger than the object. A convex mirror curves outward like the outside of a bowl. Light is reflected so that the image appears smaller than the object.

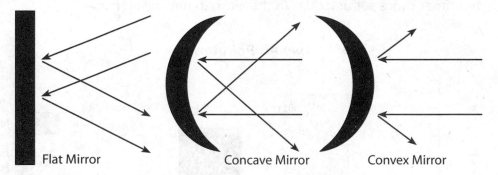

Flat Mirror   Concave Mirror   Convex Mirror

15. Which mirror reflects light so that it does not spread out?

(1) flat mirror

(2) concave mirror

(3) convex mirror

(4) both flat and concave mirrors

16. What is the main idea of the paragraph and diagram?

(1) There are three types of mirrors.

(2) The image in a flat mirror appears the same size as the object.

(3) No matter what type of mirror is used, an image appears smaller than the object.

(4) Concave mirrors are curved.

17. Mary bought a round, concave mirror for her dresser. Which of the following will she see when she looks into the mirror?

(1) her face larger than it really is

(2) her face smaller than it really is

(3) her face the same size as it really is

(4) nothing at all because concave mirrors do not make reflections

Questions 18–20 are based on the following line graph.

Changes of State

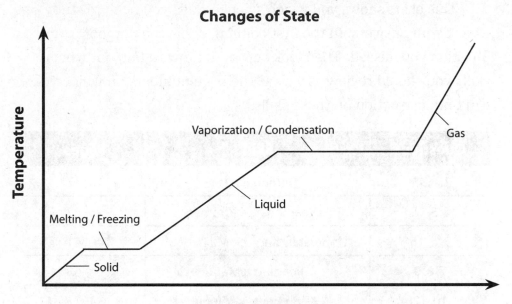

18. According to the graph, heat causes a substance to change from a
    (1) solid to a liquid.
    (2) gas to a solid.
    (3) liquid to a solid.
    (4) gas to a liquid.

19. According to the graph, which of the following substances would have the greatest thermal energy?
    (1) ice cubes
    (2) ice water
    (3) warm water
    (4) boiling water

20. Which instrument could you use to measure the thermal energy of a substance?
    (1) scale
    (2) thermometer
    (3) calculator
    (4) tape measure

When you finish the *Science Pretest*, check your answers on page 158. Then look at the chart on page 10.

Skills Preview Chart

This chart shows you which science skills you need to study. Check your answers. In the first column, circle the number of any question you missed. Then look across the row to find out which skills you should review as well as the page numbers on which you can find instruction on those skills.

Questions	Skill	Pages
1, 4, 11	Finding Facts	16–17
2, 5, 15	Reading a Diagram	26–27
3, 6, 16	Understanding the Main Idea	40–41
7, 8, 9	Reading a Map	64–65
10, 12	Using Context Clues	74–75
13	Understanding Cause and Effect	98–99
14	Reading Tables and Bar Graphs	104–105
18, 19	Reading Line Graphs	120–121
20	Drawing Conclusions	130–131
17	Predicting Outcomes	136–137

Unit 1   Life Science

In this unit you will learn about

- scientific methods
- the cell
- the circulatory and nervous systems
- bones and muscles
- bacteria and viruses
- extinction
- plants
- the environment

hypothesis

bacteria

antibodies

cortex

Life science is the study of animals, plants, and other living things in our environment—the land, water, and air.

Name two living things you see around you in your environment.

_____

_____

Life science is also the study of the human body and how it works. Life science can give us tips about how to stay healthy.

What is something you do to stay healthy?

_____

_____

Have you ever eaten a home-grown tomato, picked right off the plant? If you have, you know that a home-grown tomato is much tastier than a supermarket tomato. A home-grown tomato is red, juicy, firm, and very tasty. In contrast, a supermarket tomato is often tough and mealy with a mild taste.

Supermarket tomatoes are bred to be firm so they will survive long truck hauls. They are also picked green, before they become sweet, so they will stay fresh for weeks.

Can we develop a supermarket tomato that tastes like a home-grown tomato? A scientist who asks such a question has ways to find the answer. These ways are called **scientific methods**. Scientific methods are ways of getting information and testing ideas. They give scientists a logical way to solve problems.

A first step can be to **observe**. This means scientists look carefully at a problem. In the case of the tomato, plant scientists studied home-grown and supermarket tomatoes. They observed how these tomatoes are the same and how they are different. They observed what makes home-grown tomatoes taste so good. They observed what features of supermarket tomatoes make them easy to ship and store.

scientific methods
processes for getting information and testing ideas

observe
watch and read to gather information about something

Tomatoes fresh from the garden taste better than the ones that have been hauled a long distance to the supermarket.

Once scientists observe things, they state the problem as a question. For the plant scientists, the question might be: How can we develop a tomato that can be picked when ripe yet keep for weeks?

Next, scientists might try to guess the answer to the question. This guess is not a wild guess. Instead, it is a careful guess, based on knowledge and experience. It is called a **hypothesis**. Scientists working on the tomato problem came up with this hypothesis: There is a **gene** that causes the tomato to get soft when it is ripe. If the gene is reversed, or inserted backward, in new tomato plants, then the tomatoes will ripen without softening.

Another scientific method is to **experiment**. Experiments must be designed to test whether the hypothesis is true. To test the backwards-gene hypothesis, scientists found the soft-when-ripe gene. Then, they inserted it backwards into a group of tomato plants. They also grew other tomato plants without the reversed gene. Scientists performed these experiments on many varieties of tomatoes.

hypothesis
a careful guess about the answer to a question

gene
a microscopic part of a living thing that tells the living thing how to develop

experiment
a method used to test a hypothesis

Scientists studied home-grown tomatoes.

conclusion
the decision on whether a hypothesis is supported by evidence

After an experiment, scientists look at their results. Sometimes the results support the hypothesis. Sometimes they do not. The results of an experiment are stated in a **conclusion**. The conclusions scientists make often lead to new observations and experiments. For example, the results of the experiment on tomatoes supported the scientists' hypothesis. They could use reversed genes to develop a tomato that can be picked close to ripeness and stay firm in the supermarket.

The following chart summarizes scientific methods.

Scientific Methods	
Method	**Example**
1. Observe.	1. Home-grown tomatoes, picked when ripe, taste better than supermarket tomatoes.
2. State the problem as a question.	2. How can we develop a tomato that can be picked when ripe yet keep for weeks?
3. Make a hypothesis.	3. There is a gene that causes the tomato to get soft when it is ripe. If the gene is inserted backward into new tomato plants, then the tomatoes will ripen without softening.
4. Experiment.	4. Grow some tomatoes with the reversed gene and others without it.
5. Draw a conclusion.	5. Reversed genes can produce a tomato that can be picked when ripe and then stored.

GED Tip

Some GED Science Test questions refer to information in a chart. Before you read the details in a chart, read the title to get the main idea.

Practice

Vocabulary ■ **Write the word or words that best complete each sentence.**

1. A _____ is a way that scientists solve problems.

2. A _____ is a guess, based on experience and knowledge, about the answer to a problem.

3. To test their ideas, scientists perform _____.

conclusion

experiments

hypothesis

scientific method

Finding Facts ■ **Circle the number of the correct answer.**

4. Scientists study a problem and look at it carefully. This process is called

    (1) observation.
    (2) experimentation.
    (3) hypothesis.
    (4) conclusion.

5. After scientists experiment, they decide whether the results support their hypothesis. They draw

    (1) an experiment.
    (2) a conclusion.
    (3) a scientific method.
    (4) an observation.

Sequence ■ **Write five scientific methods used by the plant scientists in the order in which they were used.**

6. _____

    _____

    _____

    _____

    _____

Check your answers on page 159.

Finding Facts

On the GED Science Test, you may need to find facts to answer questions. When you read to find facts, you can skim, or read quickly, through the information.

As you read, look for words or numbers that stand out. Sometimes, important information in science material is printed with **dark letters** that stand out. Look at the other words that surround the words that stand out. This can give clues about the meaning of those words.

 Strategy Look for facts in the paragraph.
Ask yourself: Which facts am I looking for?
1. Look at the whole paragraph. Ask yourself: Do any words stand out?
2. Look at the words that stand out. Ask yourself: What do these words tell me?

Exercise 1: Skim the following paragraph. Find the word that stands out and write it down. Then write the meaning of that word.

Have you ever heard someone say he or she has a theory about something? Scientists have theories, too. In everyday life, a theory is a guess about anything. This is not the case in science. In science, a **theory** is an explanation that is based on many experiments and observations. No experiment or observation can conflict with a theory. If it does, then the theory is changed.

word: _____

meaning: _____

_____

Look for lists when you skim through paragraphs. Lists help organize facts so that they are easier to find. Also, pay attention to numbers and capital letters as you read. These parts of the paragraph usually contain facts.

 Strategy Look for facts in the paragraph.
Ask yourself: How are the facts organized?
1. Look for lists. Lists organize facts.
2. Look for numbers and capital letters.

Exercise 2: Skim the paragraph. Then answer the questions.

Sunscreens are creams that help protect the skin from the sun's harmful rays. When you choose a sunscreen, look at the **SPF**, or Sun Protection Factor. The SPF tells how strong the protection of the sunscreen is against certain types of rays. Look for an SPF of 15 or higher. Many people think that putting on sunscreen is all they need to do, but using sunscreen is not enough. Here are ways to protect your skin from the sun.

- Slip on a shirt.
- Slop on sunscreen.
- Slap on a hat.

1. What are three ways to protect skin from the sun?

_____

_____

2. What is SPF? _____

_____

3. What number SPF should you look for? _____

Check your answers on page 159.

cell
the smallest part of a living thing that is living

bacteria
(plural of bacterium) tiny one-celled organisms

yeast
a kind of fungus that is made of one cell

microscope
an instrument that makes very small things look bigger

What do you see when you look at your hand? You might notice the skin or the wrinkles in the skin. Maybe you see blood vessels under the skin. What you will not see are the many tiny **cells** that make up your hand.

All living things are made of cells. Some living things are made of only one cell. **Bacteria** are made of one cell. **Yeasts** are also made of one cell.

What do cells look like? You cannot see the cells in your hand just by looking at it. Most cells are very small. You need the help of a **microscope** to see cells.

Key Parts of a Cell

- Cell membrane—outer covering of the cell

- Cytoplasm—jellylike material inside a cell

- Nucleus—control center of the cell

This is a diagram of a typical human cell.

A Human Cell

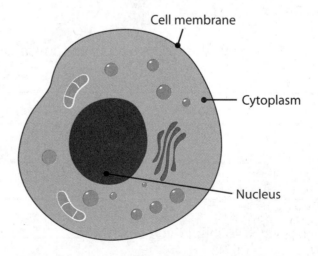

Some of the cells in your body look like this cell.

The skin on your hand is made of skin cells. The bones of your hand are made of bone cells. The muscles in your hand are made of muscle cells. The blood in the vessels in your hand is made of blood cells. In fact, your hand, like the rest of your body, is made of all different kinds of cells.

Your body is made of many cells—about 100 trillion. Animals and plants are also made of many cells. In living things made of many cells, different cells do different jobs. These cells are called **specialized cells**.

Specialized cells have parts that help them carry out specific jobs. For example, a **nerve cell** has long fingerlike parts that help it send and receive messages. Your brain is made of nerve cells. They help you see, hear, smell, taste, and touch things. By using a microscope, you can see what different types of specialized cells look like.

specialized cell
a cell that has parts that help it carry out a specific job

nerve cell
a cell with long fingerlike parts that help it send and receive messages

Specialized Cells

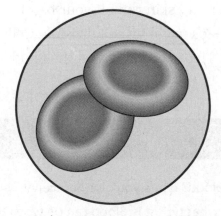
Red blood cells

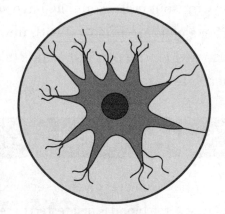
Nerve cell

Red blood cells and nerve cells are specialized cells. They have different parts and different jobs.

parent cell
a cell that splits into two
new cells

cell division
a process that happens
when one cell splits into
two new cells

organism
any form of animal or
plant life

Cells come from other cells. A **parent cell** splits into two new cells. This process is called **cell division**. Cell division is a form of reproduction for one-celled **organisms**. As you can see in the diagram, the new cells look exactly like the parent cell.

Cell Division

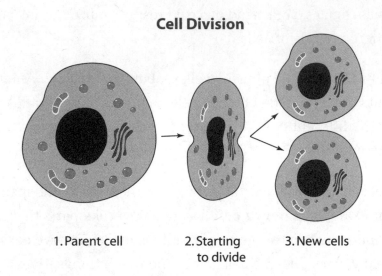

1. Parent cell   2. Starting   3. New cells
                     to divide

A parent cell splits into two new cells during cell division. Cells divide this way when a cut on the hand heals.

When cells divide in many-celled organisms, they replace old or worn-out cells. You may have seen dry skin rub off your hand. These cells are replaced with new ones made by cell division. Cells also divide to repair damaged cells. This is how a cut on the hand heals, for example.

▶ **GED Tip**

Some GED Science Test questions are based on a paragraph. As you read, think about the facts in the paragraph and how they fit together.

Blood Cells and Blood Types

Have you ever given or received blood? When you give or receive blood, the blood is first tested to see what type it is. Blood can be type O, A, B, or AB.

There is sugar on the surface of red blood cells. The type of sugar on the cell membrane of blood cells determines what type of blood you have.

Practice

Vocabulary ■ Write the word or words that best complete each sentence.

1. The different types of cells that make up many-celled organisms are called _____.

2. When a cell splits into two new cells, the process is called _____.

3. An instrument that can be used to see cells is a _____.

bacterium

cell division

specialized cells

microscope

Finding Facts ■ Circle the number of the correct answer.

4. Which kind of cell has long fingerlike structures that help it send and receive messages?

   (1) blood cell
   (2) bone cell
   (3) muscle cell
   (4) nerve cell

5. Which of these is a one-celled living thing?

   (1) animal
   (2) bacterium
   (3) human
   (4) plant

Finding Facts ■ Read the question. Then write your answer.

6. What substance on the surface of red blood cells determines blood type?

   _____

Check your answers on page 160.

Your heart is a pump. It is about as big as your fist and weighs less than a pound. The heart is located in the middle of the chest, between the lungs. It beats about 70 times a minute, more than 100,000 times a day. It pumps blood all day, every day. The heart is always on duty. When you run, it speeds up. When you rest, it slows down.

Adults have between four and five quarts of blood in their bodies. All of the blood passes through the heart once every minute. The heart pumps blood to every part of the body through tubes called **blood vessels**. There are about 100,000 miles of blood vessels in the body. If you put all your blood vessels in a line, they would stretch around the world four times.

There are three main kinds of blood vessels: **arteries**, **veins**, and **capillaries**. The arteries carry blood away from the heart. The veins carry blood back to the heart. The capillaries deliver blood to the body's cells. The heart, blood, and blood vessels work together. The blood is always circulating, or moving around the body. For this reason, the heart and the blood vessels are called the **circulatory system**.

In order for the body to live, all the cells in the body need oxygen. Cells get oxygen and nutrients, or food, from the blood. Arteries send fresh blood that is rich in oxygen all around the body. The arteries don't deliver the fresh blood directly to the cells. The arteries are like large pipes. They deliver the blood to smaller pipes, the capillaries. A capillary is very small. Only one blood cell at a time can pass through its walls. Each blood cell then takes oxygen and food to the body's cells.

blood vessels
tubes that blood flows through

arteries
blood vessels that carry blood away from the heart

veins
blood vessels that carry blood to the heart

capillaries
tiny blood vessels that deliver blood to the body's cells

circulatory system
the heart and all the blood vessels

Body cells produce waste. Blood cells return this waste to the capillaries. Capillaries pass the "used" blood on to the veins. The veins then take the used blood to the liver, lungs, and kidneys. These organs clean the waste from the blood. The cleaned blood then returns to the heart.

Next, the heart pumps the cleaned blood to the lungs. Here the blood receives a fresh supply of oxygen. From the lungs, the "fresh" blood goes back to the heart. The heart pumps the fresh blood to the arteries and around the body again.

The Circulatory System

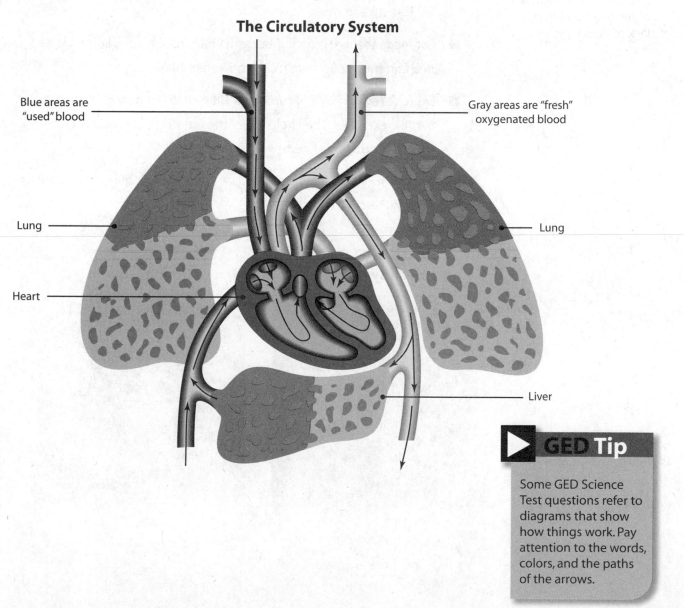

Blue areas are "used" blood

Gray areas are "fresh" oxygenated blood

Lung

Lung

Heart

Liver

▶ **GED Tip**

Some GED Science Test questions refer to diagrams that show how things work. Pay attention to the words, colors, and the paths of the arrows.

coronary artery disease
a disease that causes the arteries to become blocked

cholesterol
a type of fat found in the human body and in foods made from animals

heart attack
a condition in which part of the heart dies from lack of blood

Diseases of the heart and circulatory system can be deadly. The most common heart disease is **coronary artery disease**. In this disease, the blood vessels leading to the heart become blocked with fat and **cholesterol**. As a result, the heart does not get enough blood. Part of the heart can die. When this happens, a person has a **heart attack**. Each year, about 1,500,000 Americans have heart attacks. About one third of them die.

Reducing the Risk of Heart Disease

- Avoid foods that are high in fat and cholesterol. Eat fewer eggs. Eat less cheese, meat, milk, and ice cream.

- Eat foods that are high in fiber and vitamins. Eat whole-grain bread, rice, and pasta. Eat more fruits and vegetables.

- Exercise regularly. Walk, run, swim, bike, skate, or dance. These are aerobic exercises. They help lower the amount of fat and cholesterol in the blood. They also strengthen the heart.

- Do not smoke. Heavy smokers have twice the risk of heart disease as nonsmokers.

Healthy artery   **Clogged artery**

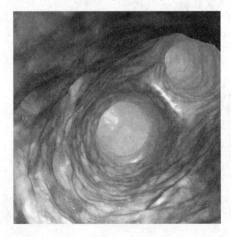
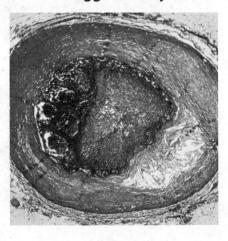

The artery on the left is healthy. The artery on the right is clogged with fat.

Practice

Vocabulary ■ Write the word that best completes each sentence.

1. Blood vessels that carry blood to the heart are called

    _____.

2. Blood vessels that carry blood away from the heart are called

    _____.

3. Tiny blood vessels that deliver fresh blood to the cells are called

    _____.

arteries

capillaries

cholesterol

veins

Finding Facts ■ Circle the number of the correct answer.

4. Where does blood go after it passes through the arteries?

    (1) to the capillaries

    (2) to the veins

    (3) to the heart

    (4) to the liver, lungs, and kidneys

5. Where does "fresh" blood go from the lungs?

    (1) to the capillaries

    (2) to the veins

    (3) to the heart

    (4) to the liver, lungs, and kidneys

Finding Facts ■ Write your answer below.

6. List two things people can do to lower their risk of getting coronary
    artery disease.

    _____

    _____

    _____

Check your answers on page 160.

Reading a Diagram

Diagrams are drawings that can help you understand science. One type of diagram is a drawing that shows the parts of something. Other types of diagrams show how something works. Diagrams usually have labels, or words, that explain what you are looking at.

When you look at a diagram, first figure out the main idea, or general subject, of the diagram. The title of the diagram often states the main idea. For example, the main idea of the diagram on the left is the nervous system.

The Nervous System

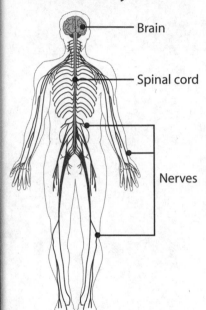

Brain

Spinal cord

Nerves

> ▶ **Strategy** Look at the diagram. Ask yourself: What is the main idea?
>
> 1. Look at the title or caption. It tells you the main idea of the diagram.
>
> 2. Look at the whole drawing. Ask yourself: What is this drawing about?

Exercise 1: Look at the diagram below. Then write the main idea.

_____

_____

Muscles of the Upper Arm

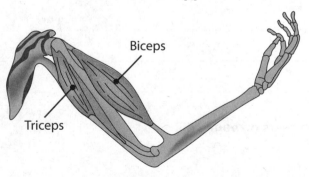

Biceps

Triceps

Once you understand the main idea, look carefully at the details. Figure out what the parts of the diagram show. Often there are labels that tell what the parts are or what they do. In the diagram of the nervous system on page 26, the labels are *brain, spinal cord, nerves*. Each label shows you an important detail.

 Strategy Look at the diagram. Ask yourself: What are the important details of this diagram?

1. Look at the parts of the diagram. Ask yourself: What do these parts show?

2. Read the labels. They tell you the important details.

Exercise 2: Look at the diagram. It shows what different areas of the brain do. Write two details that the diagram shows.

_____

_____

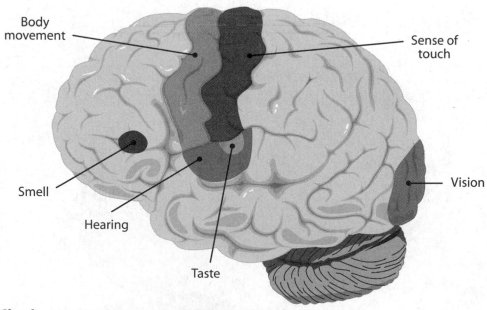

Areas of the Brain

Check your answers on page 160.

brain
the part of the body that controls all the body's activities

cerebrum
the part of the brain in which thinking occurs

cortex
the part of the cerebrum that stores information

cerebellum
the part of the brain that coordinates muscle activities

brain stem
the part of the brain that controls automatic life processes

What part of you controls automatic activities like blinking and breathing? What body part makes sense of what you see, hear, touch, taste, and smell? What body part gives instructions to your muscles when you walk down the street? What part of you thinks thoughts and feels feelings? If you answered the **brain**, you are right.

Your brain is the most complex part of your body. It weighs only three pounds, yet it controls everything you do. The brain is divided into three major parts.

1. The largest part, the **cerebrum**, is responsible for thinking and deciding to act. It is divided into a left side and a right side. The outer layer of the cerebrum, the **cortex**, is where most information is stored.

2. The **cerebellum** is one-eighth the size of the cerebrum. It makes the muscles work together.

3. The **brain stem** controls basic life functions like breathing and heartbeat. It also controls how alert you are.

The Human Brain

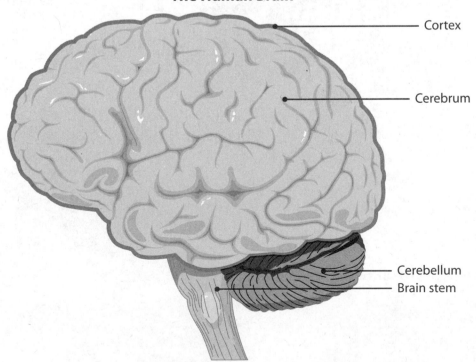

Cortex

Cerebrum

Cerebellum
Brain stem

Your brain is the control center for sending and processing information. The **nervous system** consists of the brain and the nerves that go through the **spinal cord**. From the spinal cord, nerves go to every part of the body.

The Nervous System

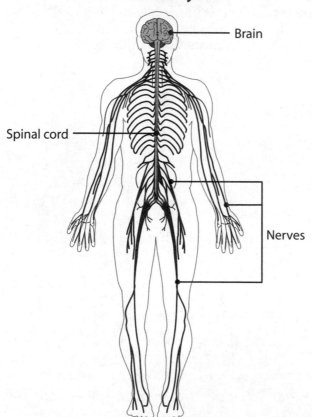

Brain

Spinal cord

Nerves

nervous system
all the nerves in the body and the brain

spinal cord
a part of the body that allows messages to travel between the brain and the body

reflex
a quick response caused by nerves in the spinal cord

Nerves throughout your body carry information to and from the spinal cord and brain. For example, nerves in your fingertips send information about what you touch to the spinal cord and brain. Nerves that go to **organs** such as the kidneys and heart are also part of the nervous system.

organ
part of the body, such as the heart or skin, that does a particular job

The brain and the spinal cord receive, process, and respond to information. For example, you might see a dog, recognize it, and pat its head. This information is processed in your brain. On the other hand, you might touch a hot pot handle and drop the pot. This is an automatic **reflex**. Nerves in your spinal cord make you drop the pot.

The nervous system can be damaged. If the brain, nerves, or spinal cord is hurt, the results can be serious. When part of the brain is damaged, a person can lose one or more abilities.

For example, a person who has an accident causing injury to the brain may not be able to remember anything that happened before the accident. If a person's spinal cord is damaged enough, that person cannot feel and move below the point of the injury.

Spinal Cord Injuries

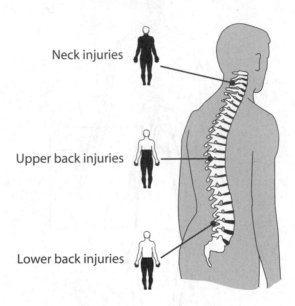

Neck injuries

Upper back injuries

Lower back injuries

The diagram shows how the body is affected when the spinal cord is damaged in different places. The shaded areas on the small pictures show where a person cannot move and feel.

Fortunately, the nervous system is well protected. The brain is protected by the hard skull. The spinal cord is inside a strong bony covering. Both the brain and spinal cord are surrounded by a liquid that softens shocks. Even the major nerves are well covered. Most of them are below layers of muscle or behind bony areas like the eye socket.

GED Tip

Some GED Science Test questions refer to diagrams with different parts labeled. Look at how the lines label the parts of the diagram.

Wearing a Helmet

Because brain damage is so serious, many cities and states require people to protect themselves when taking part in certain activities. For example, motorcycle riders may be required to wear helmets. People may also be required to wear helmets when riding bikes.

Practice

Vocabulary ■ **Write the word or words that best complete each sentence.**

1. The control center of the nervous system is the

_____ .

2. The _____ is the part of the brain that controls basic activities like breathing and heartbeat.

3. Thinking takes place in the _____ .

4. The _____ makes sure the muscles are working together.

brain

cerebellum

cerebrum

reflex

brain stem

Finding Facts ■ **Circle the number of the correct answer.**

5. For which of the following activities is the brain responsible?

(1) providing energy for the cells in the body

(2) pumping blood throughout the body

(3) making sense of what you hear, see, touch, taste, and smell

(4) removing waste from the blood

6. What is a reflex?

(1) a complicated thought

(2) the result of a brain injury

(3) part of the spinal cord

(4) a quick response

7. Which of the following protects the nervous system?

(1) the heart

(2) the bones

(3) the nerves

(4) the eyes

Check your answers on page 160.

The Skeleton

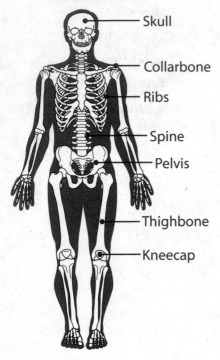

- Skull
- Collarbone
- Ribs
- Spine
- Pelvis
- Thighbone
- Kneecap

marrow
the material inside bones that produces blood cells

osteoporosis
a disease that causes weak bones

Your bones support and protect your body. The 206 bones of the human skeleton carry the body's weight. With the help of muscles, your skeleton allows you to move. It also protects your organs. The skull covers the brain. The ribs protect the heart and lungs.

Bones are very light. In fact, the skeleton of a 160-pound person weighs only about 29 pounds. Bones are light for two reasons. First, they have many tiny holes that are pathways for blood vessels and nerves. Second, some bones, such as the long bones of the arms and legs, are like tubes. The tubes contain **marrow**, which is light.

Despite their light weight, bones are extremely strong. Your skeleton can hold up more than five times its weight. Pound for pound, for example, your thighbones are stronger than reinforced concrete. This strength comes from the materials that make up bone. About half of a bone's weight is made up of hard minerals like calcium and phosphorus. A quarter of its weight is made up of a protein fiber. The rest of a bone is mostly water. The minerals and fiber are held together firmly, making bones strong.

Keeping Bones Strong

Osteoporosis is a disease of the bones. The bones become weak and may break. Osteoporosis affects millions of men and women. Here are some ways to prevent it.

- Eat foods that are high in calcium, such as milk, yogurt, cheese, broccoli, and spinach. Some cereals, orange juices, and breads have calcium added to them.

- Exercise regularly. Exercises that put weight on the bones, such as walking and lifting weights, are best.

- Do not smoke or drink alcohol. These activities weaken the bones.

Your bones are strong, but to hold you up they need help. **Ligaments** hold the bones together. Muscles move bones or keep them in place. Muscles are attached to bones by **tendons**. Bones, ligaments, muscles, and tendons work together to help you stand and move.

Muscles move bones by pulling them. Because muscles can only pull, not push, they work in pairs. One muscle of the pair contracts, or shortens, pulling the bone toward it. At the same time, the other muscle of the pair relaxes, allowing the bone to move. To make the bone move the other way, the first muscle relaxes and the second one contracts.

ligaments
tough, elastic bands that hold two bones together

tendons
tough, elastic bands that attach muscles to bones

How Muscles Work

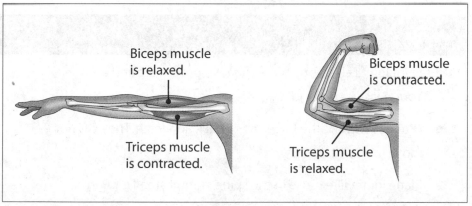

Biceps muscle is relaxed.

Triceps muscle is contracted.

Biceps muscle is contracted.

Triceps muscle is relaxed.

For example, when you bend your arm at the elbow, you contract the biceps muscle on the front of the upper arm. The triceps muscle on the back of the upper arm relaxes. When you straighten your arm, the triceps contracts and the biceps relaxes.

The Muscles

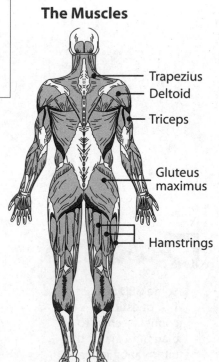

Trapezius
Deltoid

Triceps

Gluteus maximus

Hamstrings

Your muscles and bones move in thousands of ways. The body is flexible because of the way bones are linked together. The human body contains several types of **joints** that permit different kinds and ranges of movement.

Some joints allow no movement at all. These joints are called fixed joints. The joints between the bones in the skull are fixed joints.

Some joints allow only a little movement. They are called slightly movable joints. The joints between the bones of the spine are slightly movable.

Freely movable joints allow movement in one or more directions. The joints in your elbow, knee, hip, and shoulder are examples of freely movable joints. Read the chart below.

Joints

- Pivot joints allow one bone to move or turn around another. Your neck is a pivot joint.

- Hinge joints, like the knee, move back and forth. They cannot move from side to side.

- Plane joints, like those in the spine, permit small gliding movements.

- Ball-and-socket joints, like the shoulder and hip, move freely in almost all directions.

- Saddle joints, like those of the ankle and thumb, allow movement in two directions at right angles.

Five Types of Joints

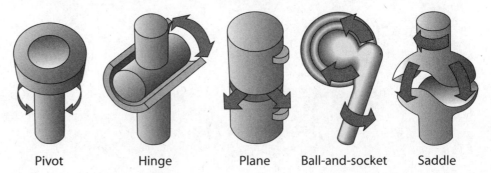

Pivot   Hinge   Plane   Ball-and-socket   Saddle

Practice

Vocabulary ■Write the word or words that best complete each sentence.

1. Blood cells are produced by material called _____ in the center of certain bones.

2. A _____ holds two bones together; the place where bones are linked is called a _____.

3. A _____ connects a muscle to a bone.

joint

ligament

marrow

skeleton

tendon

Finding Facts ■Circle the number of the correct answer.

4. Why are bones so light?

   (1) Bones are like tubes.
   (2) Bones are made almost entirely of calcium.
   (3) Bones are made of a protein fiber.
   (4) Bones contain water.

5. Why are bones so strong?

   (1) The marrow in bones makes them strong.
   (2) Minerals and protein fiber are held together firmly.
   (3) Tendons attaching them to muscles make them strong.
   (4) Ligaments that connect bones make them strong.

Reading a Diagram ■Read the question. Then write your answer.

6. A fracture is a break in a bone. This diagram shows fractures in the bones of the lower arm. What is the main idea of the diagram?

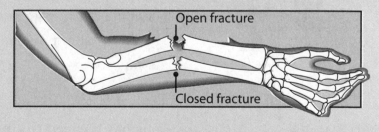
Open fracture

Closed fracture

_____

_____

_____

Check your answers on page 161.

Lesson 6   Bacteria and Viruses

bacteria
tiny one-celled organisms

virus
a tiny part of a cell that can copy itself only by using living cells

Shapes of Bacteria

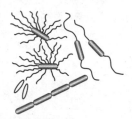

Bacilli

Spirilla

Cocci

toxins
poisonous chemicals

All around you, there are tiny organisms called **bacteria** and cell parts called **viruses**. They are in the air we breathe, the water we drink, and the food we eat. Although most bacteria and viruses do not harm people, many can cause disease. Disease-causing bacteria and viruses are often called germs.

Bacteria are divided by shape into three main groups that are listed in the chart below.

Shapes of Bacteria

- bacilli—rod-shaped bacteria

- spirilla—coiled bacteria

- cocci—sphere-shaped bacteria

Different kinds of bacteria cause different diseases. For example, bacilli cause salmonella food poisoning, tuberculosis, and whooping cough. Spirilla cause Lyme disease and syphilis. Cocci cause strep infections and acne.

Bacteria cause disease in two ways. Some bacteria cause disease by increasing quickly in the body. They reproduce by dividing in two over and over again. In a few hours, one or two bacteria can reproduce into millions. This is how tuberculosis is caused.

The second way bacteria cause disease is by giving off chemicals called **toxins**. The toxins damage parts of the body, sometimes causing death. Botulism is a deadly type of food poisoning caused by a toxin.

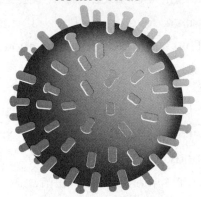

Round virus

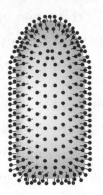

Bullet–shaped virus

Viruses have different shapes.

Viruses are much smaller than bacteria. However, like bacteria, viruses have different shapes. Viruses that cause the common cold, for example, are round. The virus that causes rabies is shaped like a bullet. Other viruses are shaped like rods or cubes. All viruses have an outer layer, called a protein coat, which protects the inside of the virus.

Viruses cause disease by taking over living cells. A virus enters a living cell and uses it to make copies of the virus. This damages or destroys the cell. As each new virus enters a new cell, the disease spreads through the body. Most viruses attack a particular type of cell. For example, the flu virus attacks the cells in the lining of the lungs, throat, and nose.

The body can fight disease-causing bacteria and viruses. When a germ enters the body, the **immune system** responds. The immune system produces **antibodies** specifically for that type of germ. These antibodies destroy the germs. The immune system also sends **white blood cells** to help destroy the germs.

If you get a disease caused by bacteria, your doctor may give you an **antibiotic**. Antibiotics stop bacteria from reproducing. However, bacteria can change over time into new kinds of bacteria. When this happens, antibiotics may no longer be able to stop these bacteria. Bacteria can change when people take antibiotics that they do not need. Bacteria can also change when a person who is sick does not finish all of the antibiotic prescribed by a doctor.

immune system
the body's system for fighting disease-causing germs

antibodies
chemicals made by the immune system to destroy a particular type of germ

white blood cells
cells that can surround and kill germs

antibiotic
a drug that stops the growth and reproduction of bacteria

GED Tip

Some GED Science Test questions ask you to find information in paragraphs. Skim the paragraphs to quickly find the information.

Tuberculosis was once a deadly disease. Once scientists developed antibiotics, this disease could be cured. However, new forms of the tuberculosis bacteria have developed. These forms are not affected by many antibiotics. That's one reason tuberculosis is again spreading in the United States.

Antibiotics do not work against viruses. There are few drugs that can cure diseases caused by viruses. But we have many drugs that relieve symptoms while your body fights the virus. For example, a cold tablet helps stop your sneezing and runny nose. It does not attack the cold virus.

Immunization can protect against many types of disease-causing bacteria and viruses. **Vaccines**, which use a weakened or dead form of the germ, are given by injection or by mouth. They cause the immune system to make antibodies. If the immunized person later gets that germ, he or she already has antibodies to fight it. Children are usually immunized against several diseases. As a result, most children in the United States no longer get the illnesses in the chart below.

immunization
the process of creating resistance to particular germs

vaccines
substances that make the body produce antibodies against a particular germ

Childhood Immunizations

- polio
- diphtheria, pertussis (whooping cough), tetanus (DPT)
- measles, mumps, rubella (MMR)
- chicken pox

Unfortunately, scientists have not yet been able to develop vaccines for many other diseases. Colds, for example, are caused by more than 100 different types of viruses. If we had a vaccine for one or two of those viruses, it would not protect against the other types of viruses. Also, scientists have not developed a vaccine for HIV, the virus that leads to AIDS. The only defense against these viruses is to prevent becoming infected.

Practice

Vocabulary ■ Write the word or words that best complete each sentence.

1. Disease-causing _____ and _____ are often called germs.

2. Some bacteria cause disease by giving off _____ , or poisons.

3. _____ are chemicals produced by the immune system to fight particular germs.

antibiotics
antibodies
bacteria
toxins
viruses

Finding Facts ■ Circle the number of the correct answer.

4. What are antibiotics used for?

   (1) to give immunity against a disease
   (2) to relieve the symptoms of a disease
   (3) to stop the growth and spread of bacteria
   (4) to stop the growth and spread of viruses

Reading a Diagram ■ The diagram shows how a virus uses a cell to make new viruses. Read each question. Then write your answer.

5. What happens to the cell at the end of this process?

   _____

6. What detail in the diagram helped you answer this question?

   _____

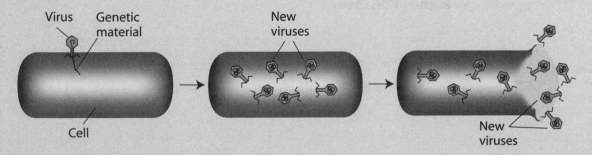

Virus   Genetic material   New viruses   New viruses   Cell

Check your answers on page 161.

Understanding the Main Idea

All the sentences in a paragraph usually relate to each other in some way. In other words, there is one main idea. The main idea is often stated in one of the sentences of the paragraph. It is often in the first or last sentence. Knowing the main idea will help you understand the whole paragraph better.

 Strategy Find the sentence that tells the main idea. Ask yourself: What is the paragraph about?

1. Read the whole paragraph.

2. Find one sentence that sums up all the other sentences. Ask yourself: Which sentence tells what the whole paragraph is about?

Exercise 1: Read the paragraph. Underline the third sentence. This sentence gives the main idea.

If someone asked you to list the important organs of the human body, you would probably mention the brain, the heart, and maybe the lungs or kidneys. You probably wouldn't mention the skin. But the skin is as important as the other organs. You have about 18 square feet of skin. Your skin is the largest organ of your body. Without it, you would die.

Exercise 2: Read the paragraph. Underline the sentence that gives the main idea.

There are three kinds of blood vessels: arteries, veins, and capillaries. The arteries carry the blood away from the heart. The veins carry blood back to the heart. The capillaries deliver the blood to the body's cells.

Sometimes the main idea isn't directly stated in one of the sentences of a paragraph. Instead, it is suggested, or implied. The sentences of the paragraph are still related. You can use the sentences to figure out what the main idea is.

Read the paragraph. What is the main idea?

Bacteria and viruses are in the air we breathe. They're in the water we drink. They're in the food we eat. They're on the outside and on the inside of our bodies.

The paragraph mentions several different places where you can find bacteria and viruses. The main idea of the paragraph is that bacteria and viruses are found almost everywhere.

 Strategy Reread the paragraph for details.
Ask yourself: How are the details related?

1. Think about how all the sentences are related to each other. This is your clue.

2. Decide what main idea all the sentences support. This is the implied main idea.

Exercise 3: Read the paragraph. Then write the main idea.

Skin helps keep the body at the right temperature. For example, when you're warm, sweating cools you off. Skin also protects the body from dirt and germs. When you get a cut, you lose some of that protection.

_____

_____

Check your answers on page 161.

Lesson 7   Extinction

hairy mammoths
large, fur-covered, elephant-like animals that no longer exist

herbivores
animals that eat only plants

prehistoric
existing millions of years ago

extinct
no longer in existence

fossils
the remains of once-living things

Until about 10,000 years ago, **hairy mammoths** lived in northern Asia, Europe, and North America. Their long fur protected them from the cold. Larger than elephants, hairy mammoths grew to 12 feet tall. They had tusks up to 16 feet long. They weighed as much as three or four cars. Despite their large size, mammoths did not hunt animals. They were **herbivores**.

Mammoths had several natural enemies. Saber-toothed tigers and wolves hunted mammoths. However, their most dangerous enemies were humans. **Prehistoric** humans hunted mammoths for their meat, fur, and bones.

Another problem for mammoths was the weather. Over time, the earth's climate got colder. Mammoths' fur could not protect them from the colder weather. Eventually, hairy mammoths became **extinct**.

People today find many mammoth **fossils**. Most fossils are bones or tusks. More than 50,000 mammoth tusks have been found. Even whole animals, frozen solid, have been found in the Arctic.

Hairy mammoths once lived in northern Asia, Europe, and North America.

Mammoths are not the only living things that have died out. Many species have become extinct. Dinosaurs roamed the earth for millions of years but are all extinct today. Many kinds of plants are extinct, too. The saber-toothed tigers, who hunted the mammoths, are also extinct.

Tigers of many kinds have existed for at least a million years. Tigers once lived in Europe, the Americas, and Asia. Just 100 years ago, there were eight breeds, or kinds, of tigers. Now there are only five. The other breeds are extinct.

Tigers are now found in the wild only in Asia. In 1900, in India alone there were probably 40,000 Bengal tigers. Now there are fewer than 2,500. There are only a few hundred Sumatran tigers and a few dozen Javan tigers. In all, there may be only 4,000 or 5,000 wild tigers left.

Asia is the only part of the world where tigers still live in the wild.

Why have tigers been dying out? There are two main reasons. The first reason is that until recent years, big-game hunters traveled to India or Africa to hunt wild animals. New laws have stopped some big-game hunting. However, in the early 1900s, thousands of Bengal tigers were hunted and killed.

The second reason there are fewer tigers is human population growth. In 1891, there were about 280 million people in India. One hundred years later, there were about 834 million people there. As the human population grew, people changed the jungles to farmlands. Many small animals died because their **habitat** was destroyed. These small animals were the main food for tigers, which are **carnivores**. Lack of living space has also caused tigers to die out.

habitat
the place where an animal or plant lives

carnivores
animals that eat only meat

*The lion-tailed macaque
is being trained to live
in the wild.*

Jaguars, wild cats that are smaller than tigers, are also
carnivores. They eat animals, fish, frogs, and alligators. Jaguars live
in forests or open deserts; they can climb trees and swim. Jaguars
can be found from Arizona in the United States to Argentina in
South America. Like the tiger, the jaguar is an **endangered
species**. Because of hunting and loss of habitat, jaguars are now
rare in most areas. In Argentina, for example, fewer than 200 wild
jaguars are left.

As wild habitats have become smaller, zoos have become more
important in saving endangered species. Workers in some zoos
breed animals, like tigers, to keep them from becoming extinct.
Some zoos breed and train animals so they can learn to live in the
wild. For example, the San Diego Zoo is training an endangered
species of monkey called the lion-tailed macaque. Once the
monkeys have been trained, they will be released in a wild area
of India, their original home.

Another endangered species being bred at the San Diego Zoo
is the California condor. Condors are the largest birds on earth.
From tip to tip, their wings are 9 to 10 feet long. These birds were
almost extinct when the zoo captured the few that were left in
order to protect them. For several years, the zoo has been
breeding the condors. Several condors have been let go in the
mountains of California.

Helping one species, like the condor, sometimes helps many
other species. For example, other species of rare or endangered
plants and animals live in the same habitat as the condor. By
helping to save the condor's habitat, people also help save the
habitat for other living things. Then other plants and animals that
share the habitat have a better chance to survive.

In 1980, 224 species of plants and animals were listed as
endangered in the United States. In 2005, the number had jumped
to 988. Laws now protect many of these plants and animals and
their habitats.

Practice

Vocabulary ■ Write the word that best completes each sentence.

1. _____ species no longer exist at all.

2. _____ species may soon die out if we don't protect them.

3. _____ are animals that eat only meat.

carnivores

endangered

extinct

herbivores

Finding the Main Idea ■ Circle the number of the correct answer.

4. What is the main idea of the first paragraph on page 43?

   (1) Tigers are found all over the world.
   (2) Several tiger breeds became extinct in the last 100 years.
   (3) Tigers were once common, but now they are rare.
   (4) There are five breeds of tigers today.

5. What is the main idea of the first paragraph on page 44?

   (1) The jaguar is an endangered species.
   (2) Jaguars are carnivores.
   (3) Jaguars are found from Arizona to Argentina.
   (4) There are fewer than 200 jaguars in Argentina.

Finding Facts ■ Circle the number of the correct answer.

6. What caused mammoths to become extinct?

   (1) The animals they ate died out.
   (2) The plants they ate died out.
   (3) Saber-toothed tigers became extinct.
   (4) The earth's climate became colder.

7. What is being done to save endangered animals?

   (1) Zoos are breeding them.
   (2) Laws protect their habitats.
   (3) Laws protect them from hunters.
   (4) all of the above

Check your answers on page 161.

All animals, including humans, depend on plants to get energy. Energy is needed to live. Plants make their own food by using the energy of the sun. Animals can't get energy this way. Instead, animals eat plants and use their energy. Then, plant-eating animals are eaten by other animals. So even when you eat a steak, you are getting energy from plants. The diagram below shows how energy from the sun flows through plants to animals.

Flow of Energy

Plants give off oxygen into the air. All animals need oxygen to live. We also get wood for houses, furniture, and fuel from plants. Plants are the source of many medicines. Cotton, linen, and rayon clothing come from plants.

When you walk in a park or in the woods, you are surrounded by plants. There are more than 300,000 species of plants in the world. Some are smaller than a pinhead. Others, like giant sequoia trees, are so big that you can drive a car through a tunnel cut in the trunk. To better understand the great variety of plants, scientists divide them into groups.

- **Mosses and liverworts** are small plants that have no roots, stems, or leaves. They cover the ground like a rug.

- **Ferns** grow in shady places. They are larger than mosses. They have roots, stems, and leaves. They do not have flowers or seeds.

- **Seed plants** are plants that make seeds. They also have roots, stems, and leaves.

There are two kinds of seed plants: those that make flowers and those that do not. Most seed plants are flowering plants. Flowering plants make seeds inside a **fruit**. Seed plants, or nonflowering plants, have seeds on a **seed cone**. A pine cone is a seed cone. The pine tree's seeds are on the scales of the pine cone.

mosses and liverworts
small plants that grow in damp places

ferns
plants with roots, stems, and leaves but no seeds

seed plants
plants with roots, stems, leaves, and seeds

fruit
the part of a flowering plant that contains seeds

seed cone
a scaly part of a nonflowering plant that holds the plant's seeds

Mosses, Ferns, and Seed Plants

moss

fern

sunflower

Parts of a Flowering Seed Plant

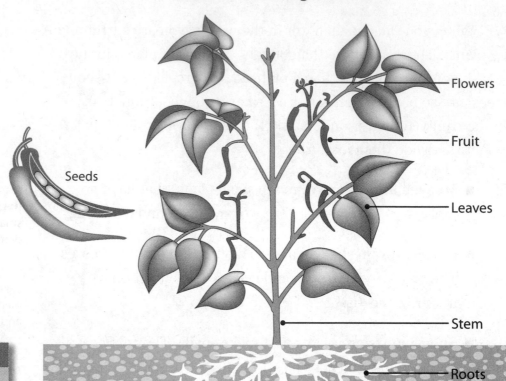

The parts of a flowering seed plant work together to keep the plant alive and to make more plants. Roots hold the plant in the ground. They take in water and minerals from the soil up into the stem. The stem supports the leaves, flowers, and fruit. The stem also carries water and minerals to the upper parts of the plant. The leaves are the plant's factory. They take in energy from the sun, make sugar, and give off oxygen into the air. The flowers of the plant produce seeds, which are protected by a fruit.

Flowering seed plants are the most common type of plants. More than half of the plant species are flowering seed plants. Most plants grown in gardens or as crops for food are flowering seed plants. Carrots and beets are the roots of flowering seed plants. Corn, peas, and lima beans are seeds. Peaches, squash, and tomatoes are fruits. Foods don't have to be sweet to be fruits. Any part of a plant that has a seed inside is a fruit.

Practice

Vocabulary ■ Write the word that best completes each sentence.

1. _____ are small plants that do not have roots, stems, or leaves.

2. _____ are plants with roots, stems, and leaves. They do not have flowers or seeds.

3. _____ plants have roots, stems, leaves, and either fruit or cones.

cones

ferns

mosses

seed

Finding the Implied Main Idea ■ Circle the number of the correct answer.

4. What is the main idea of the second paragraph on page 46?

    (1) Plants clean the air we breathe.
    (2) Wood is used to build houses and furniture.
    (3) Many fuels come from plants.
    (4) Plants have many uses.

Finding Facts ■ Circle the number of the correct answer.

5. Which part of a plant makes sugar?

    (1) roots
    (2) stem
    (3) leaves
    (4) flower

6. Which part of a plant protects seeds?

    (1) roots
    (2) fruit
    (3) flower
    (4) leaves

Check your answers on page 162.

environment
everything that surrounds us, including air, water, soil, plants, and animals

pollution
the dirtying or poisoning of the environment

algae
a very simple type of plantlike life that usually grows in water

fossil fuels
coal, oil, gas, and other fuels that formed from the bodies of living things millions of years ago

Of all the animals on the earth, people make some of the biggest changes in the **environment**. We build cities, factories, and farms. These things help support a large human population, but they can also cause harm to the environment.

Water in lakes, ponds, streams, and rivers often becomes polluted. Much of this **pollution** comes from chemical fertilizers used to grow crops. Some fertilizer is washed out of the soil by rain, and it flows into nearby lakes and streams. Fertilizer makes **algae** grow on the surface of the water. This blanket of algae blocks sunlight from reaching plants below. These plants die and decay and stop putting new oxygen in the water. Then fish start to die because there isn't enough oxygen. Eventually, most plant and animal life in a polluted pond or lake dies.

Oil and other chemical waste from factories can pollute lakes and streams. Most plants and animals cannot live in polluted water.

Types of Pollution in Our Environment

When humans harm the natural surroundings, this is called environmental pollution. The main kinds of pollution in our environment are

- Air pollution—when people burn **fossil fuels** in power plants and cars

- Soil pollution—when dirty water runs over onto land

- Water pollution—when substances like chemicals and sewage get into rivers, lakes, and oceans

Other kinds of pollution include

- Noise pollution—when machines add noise to the environment

- Radiation pollution—an invisible form of pollution that harms cells and causes cancer

- Pesticides—chemicals that are sprayed on crops that get into the air, water, and soil

Cities, suburbs, and factories add to water pollution, too. In some places, wastes from sewers run into streams and lakes. When it rains, water runs off highways and parking lots. This sometimes sends spilled oil and other chemicals into lakes and streams. Factories sometimes spill wastes into lakes and streams. These wastes can poison animals that live in the water.

The environment can also be damaged by polluting the **food chain**. For example, chemicals called pesticides can get into the food supply. A well-known example is **DDT**, a chemical that kills harmful insects. Unfortunately, DDT also kills useful insects, like ladybugs. However, DDT caused an even worse problem.

Over time people began to notice that DDT was killing animals. Bald eagles, peregrine falcons, brown pelicans, and other birds that eat fish began to disappear. Birds, fish, frogs, and other animals that eat insects had DDT in their bodies. Other animals, including humans, in the food chain had DDT in their bodies. Farm animals, close to DDT spraying, had DDT in their bodies. Traces of DDT were found in our meat and milk.

Because DDT was so harmful, it was outlawed in 1972. By 1975, DDT in humans, fish, and some birds had already begun to decrease. By the mid-1990s, the number of bald eagles had increased from 500 pairs before the ban to 5,000 pairs. The **populations** of peregrine falcons and other birds affected by DDT also grew after DDT was banned.

food chain
a cycle in which plants are eaten by animals, which are eaten by other animals

DDT
a chemical that kills insects

populations
the numbers of a group of plants or animals that live in one place

How Pesticides Enter the Food Chain

DDT

DDT

DDT

Sometimes there is disagreement over whether a chemical is safe. This is the case with **bovine growth hormone (bGH)**. Cows have their own natural bGH, but when cows are given extra bGH, they produce more milk than they normally do. However, for more than 10 years there have been arguments over whether or not giving cows bGH is safe.

Supporters of treating cows with bGH say the milk is just the same as the milk from untreated cows. Others argue against giving cows bGH. They say that treated cows get more infections of the udder. When those cows are given antibiotics to cure the infections, the antibiotics end up in the milk we drink. Antibiotics in milk can lead to diseases in humans that cannot be treated.

The U.S. government has approved the use of bGH. It was decided that the milk was safe. Most milk in the United States is made from cows that are given bGH. However, Canada and the European Union have decided to ban bGH.

Some people want the labels on milk cartons to tell whether the cows were treated with bGH.

Practice

Vocabulary ■ **Write the word or words that best completes each sentence.**

1. Coal, oil, and gas are all examples of _____.

2. The dirtying or poisoning of our surroundings is called _____.

3. _____ is a chemical that kills insects.

bGH

DDT

fossil fuels

pollution

Finding Facts ■ **Circle the number of the correct answer.**

4. Which of the following is a large source of water pollution?

   (1) chemical fertilizers
   (2) farm crops
   (3) rainwater
   (4) bGH

5. Why would farmers want to give their cows extra bGH?

   (1) to reduce the use of antibiotics
   (2) to reduce the number of udder infections
   (3) to make the milk safe to drink
   (4) to increase the amount of milk

Reading a Diagram ■ **Look at the diagram on page 51. Then answer each question.**

6. How did DDT enter the food chain?

   _____

   _____

7. What is the main idea of the diagram?

   _____

   _____

Check your answers on page 162.

Previewing Test Questions

On the GED Science Test, you will answer questions based on reading paragraphs and graphics such as diagrams, graphs, and maps.

One strategy for passing the GED Test is to preview the questions before you read the paragraph or graphic. This will help you know what to look for in the paragraph or graphic.

 Strategy Try the strategy on the example below. Use these steps.

Step 1 Read the question. What is it asking you to find out?

Step 2 Read the paragraph. Look for what the question asked you to find out.

Step 3 Answer the question.

Example

Everything in nature exists in a balance. In an ecosystem, plants, animals, and climate work together. For example, in a forest ecosystem, only a certain number of trees can grow. Trees need sunlight, space, and water. In a forest, there is only so much light, space, and water. If there are too many trees, they will compete. The stronger trees will crowd out the weaker trees until a balance is reached.

Which of the following is the main idea of the paragraph?

(1) Everything in nature exists in a balance.

(2) Strong trees can crowd out weak trees.

(3) There is only so much light in a forest.

(4) Only so many trees can grow in a forest.

In Step 1 you read the question. It asked you to find the main idea of the paragraph. In Step 2 you read the paragraph and looked for the main idea. In Step 3 you answered the question. The correct answer is (1). The main idea is stated in the first sentence of the paragraph. Choices (2), (3), and (4) are also stated in the paragraph, but they are details, not the main idea.

Practice the strategy. Use the steps you learned. Circle the number of the correct answer.

European settlers damaged the plants and animals in the Great Plains in two main ways. First, farmers plowed up the tough prairie grass and planted crops. Second, hunters killed buffalo for meat and fur. By 1900, fewer than 1,000 buffalo were left of the 60 million that had roamed the Great Plains. Fewer than 20,000 square miles of prairie were left.

1. How many buffalo lived in the Great Plains before Europeans arrived in America?

(1) 1,000
(2) 1,900
(3) 20,000
(4) 60 million

Practice the strategy with a graphic. Use the steps you learned. Circle the number of the correct answer.

2. Which of the following is the main idea of this diagram?

(1) Legs and hearts have similar veins.
(2) Blocked arteries can be repaired in the leg.
(3) A leg vein can be used to bypass a blocked artery.
(4) A blocked coronary artery is a dangerous condition.

**Using a Leg Vein
to Bypass a Blocked Artery in the Heart**

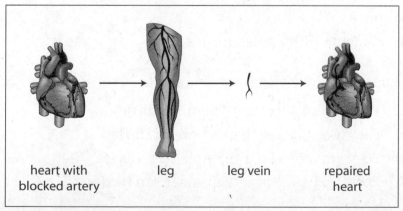

| heart with blocked artery | leg | leg vein | repaired heart |

Check your answers on page 162.

Read each paragraph and question carefully. Circle the number of the correct answer.

Questions 1–3 are based on the following paragraph.

An ecosystem can become unbalanced if a new species, or kind, of animal is released into a new area. This happened when American turtles were shipped from Florida to France. The turtles were to be sold as pets in France. Many of the American turtles escaped or were let go when they got big. The American turtles are bigger than the French turtles. In the wild, American turtles compete for food and space with the French turtles. They are driving the French turtles out of the ecosystem.

1. Which of the following is the main idea of this paragraph?

   (1) American turtles are driving out French turtles.

   (2) American and French turtles compete in the wild.

   (3) An ecosystem can become unbalanced if a new species is released in an area.

   (4) Ecosystems are balanced when species from different areas are allowed to compete.

2. Turtles were shipped from Florida to France so that they could

   (1) mate with French turtles.

   (2) be sold as pets.

   (3) drive French turtles out.

   (4) compete with French turtles.

3. Which of the following is a fact about the French turtles?

   (1) They are smaller than American turtles.

   (2) They are stronger than American turtles.

   (3) They are driving out the American turtles.

   (4) They are better pets than American turtles.

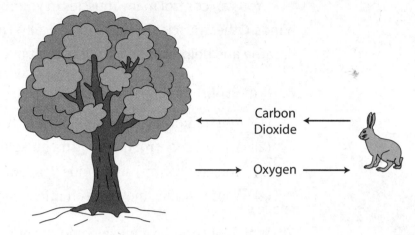

Plants make oxygen as a waste gas. Animals make carbon dioxide as a waste gas.

4. Which of these is a detail shown in the diagram?

   (1) Rabbits eat plants.
   (2) Plants give off oxygen.
   (3) Animals take in carbon dioxide.
   (4) Plants use carbon dioxide to make sugar.

5. What is the main idea of the diagram?

   (1) Rabbits breathe in the oxygen from plants.
   (2) Rabbits and plants get food from each other.
   (3) Plants and animals use each other's waste gases.
   (4) Plants take in the carbon dioxide from animals.

Questions 6–8 are based on the following paragraph.

You can control many muscles in your body, such as those in your hands. Other muscles move automatically. The muscles that help you breathe and help your heart work automatically.

6. Which of the following is the main idea of this paragraph?

(1) You do not have to think to breathe.
(2) You can control most of the muscles in the body.
(3) You can control some muscles, but not others.
(4) Your muscles move automatically.

7. Which of these is a muscle you cannot control?

(1) heart muscle
(2) calf muscle
(3) upper arm muscle
(4) thigh muscle

8. Your muscles move automatically

(1) when you walk.
(2) when you catch a ball.
(3) when you breathe.
(4) when you bend your arm.

Check your answers on page 162.

Unit 1 Skill Check-Up Chart

Check your answers. In the first column, circle the numbers of any questions that you missed. Then look across the rows to see the skills you need to review and the pages where you can find each skill.

Question	Skill	Page
2, 3, 7, 8	Finding Facts	pages 16–17
4, 5	Reading a Diagram	pages 26–27
1, 6	Understanding the Main Idea	pages 40–41

Unit 2

Earth and Space Science

In this unit you will learn about

- the atmosphere
- pollution and recycling
- earthquakes
- the solar system
- stars and constellations

constellation

atmosphere

earthquake

recycling

Earth and Space Science is the study of Earth's history and how Earth fits into the universe. It is also the study of how Earth changes over time and how we can take care of the earth.

One way to take care of the earth is to recycle. What is something that you recycle?

_____

Earth and Space Science is also the study of planets, moons, and the sun.

Name a planet besides Earth.

_____

Lesson 10   The Atmosphere

solar system
our sun and its
nine planets

atmosphere
all the air
surrounding Earth

water vapor
the moisture in the air

260°F; −280°F
a temperature of 260
degrees Fahrenheit; a
temperature of minus
280 degrees Fahrenheit

There is a wonderful difference between Earth and all the other planets. As far as we know, none of the other planets in our **solar system** can support life. Earth, unlike the other planets, is surrounded by a blanket of air called the **atmosphere**. This atmosphere is a mixture of invisible gases, mostly nitrogen, oxygen, and **water vapor**. We call this mixture *air*. Without air, our planet would be without life.

There is no life on our moon because its surface has no blanket of air around it. On the moon it is **260°F** during the day and **−280°F** at night. Life as we know it isn't possible under those extreme conditions.

In contrast, Earth's atmosphere helps keep the temperature even. The atmosphere extends thousands of miles from Earth, but it isn't the same throughout. The farther from Earth you go, the thinner the air is. Nearly three fourths of all the air in the atmosphere is within six miles of Earth's surface.

The Four Layers of Earth's Atmosphere

Thermosphere

Mesosphere

Stratosphere
(Ozone Layer)

Troposphere

Until the 1980s, people rarely worried about the atmosphere. It seemed endless and impossible to destroy. Today we know better. Scientists have studied and learned a lot about the atmosphere and how it has been affected by events on Earth.

Volcanic Eruptions

- In August 1883, Krakatoa, a **volcano** near Java, **erupted.** It sent nearly five **cubic miles** of rock and ash into the air. Darkness lasted two-and-a-half days as far as 50 miles from the volcano. Temperatures fell. Fine dust rose into the atmosphere and traveled around Earth.

- In November 1985, the Colombian volcano, Nevado del Ruiz, erupted. The explosion caused millions of tons of gas and ash to enter the atmosphere. Temperatures nearby fell 20 degrees.

volcano
an opening in the earth's crust through which melted rock is forced

erupt
explode violently

cubic mile
an area one mile long, one mile wide, and one mile high

Why do temperatures fall when the air is filled with ash and dust? The temperatures fall because the sun's light is blocked by dust. Volcanic dust causes another problem. The dust can rise high in the air. If it rises high enough, the dust will ride the winds and circle the earth for months or even years. This dust blocks sunlight.

A drop in temperature of a few degrees may not seem like much, but it is important. Dinosaurs lived on Earth for 100 million years. One theory says that the dinosaurs died because temperatures around the world fell a few degrees. Over millions of years, plants and animals get used to their climate. When the temperature changes even a few degrees, millions of plants and animals die.

Dinosaurs roamed the earth for 100 million years.

GED Tip

Some GED Science Test questions ask you to compare two graphics. Look at the labels and the direction of the arrows in diagrams to help you compare them.

People have a large impact on the atmosphere. Because of this, scientists worry about problems people could cause in Earth's climate. One problem is a **nuclear winter**. In a nuclear winter, the dust and smoke from a nuclear explosion would block out the sun's rays. Temperatures would drop. Crops and animals would freeze. People would starve.

Another problem is the **greenhouse effect**. In a greenhouse, the heat from the sun gets trapped by the glass. This heat keeps everything in the greenhouse warm. Because we are burning more and more coal and oil, more **carbon dioxide** (CO_2) is present in the atmosphere than in past years. This carbon dioxide acts like glass in a greenhouse. It lets the sun's rays in, but it doesn't let much heat out. Many scientists predict that by 2050 Earth will be about 9° F warmer than it is now.

If Earth's climate gets warmer, more water will stay in the atmosphere instead of falling as rain. This will be good in some places. For example, areas that are now too rainy or too cool for good farming will be able to grow more crops. Other areas, however, will have more dry spells and will grow fewer crops.

Many people are trying to decrease the amount of coal and oil they burn. One way is to carpool. This may help reduce the amount of warming.

The greenhouse effect: carbon dioxide in the atmosphere acts like glass in a greenhouse.

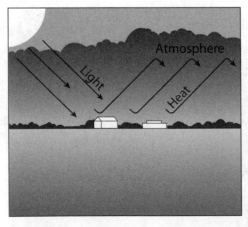

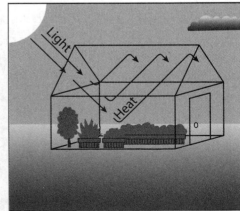

Practice

Vocabulary ■ **Write the word or words that best complete each sentence.**

1. All the air surrounding Earth is called the
   _____.

2. The drop in temperature caused by the dust and smoke from a nuclear explosion is called a _____.

3. The gases in the atmosphere let sunlight in but do not let all the heat out. This is known as the _____.

> atmosphere
> carbon dioxide
> greenhouse effect
> nuclear winter

Finding Facts ■ **Circle the number of the correct answer.**

4. Earth's atmosphere

   (1) causes the extinction of millions of plants and animals.
   (2) helps keep Earth's temperature even.
   (3) keeps sunlight from reaching Earth's surface.
   (4) extends six miles from Earth.

5. An increase in the greenhouse effect is caused by an increase in the amount of

   (1) oxygen in the atmosphere.
   (2) nitrogen in the atmosphere.
   (3) carbon dioxide in the atmosphere.
   (4) water vapor in the atmosphere.

Reading a Diagram ■ **Read the question. Then write your answer.**

6. Use the diagram on page 62 to describe how carbon dioxide gas and greenhouse glass are similar.

   _____

   _____

   _____

Check your answers on page 163.

Reading a Map

A map is a diagram that shows where places are. Sometimes maps also show features of a place. Here is a map of the United States. It also shows the weather on one day. This map is like the weather maps you see in the newspaper.

High Temperatures and Weather for January 13

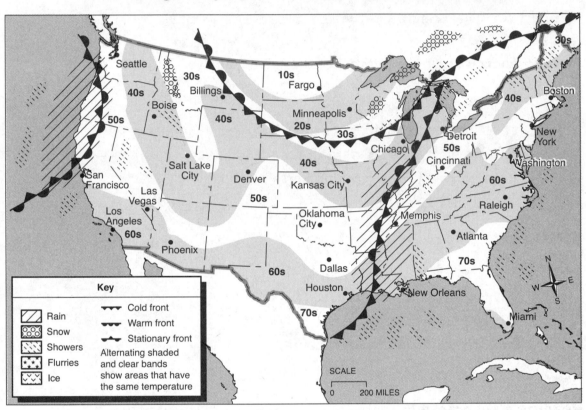

When you look at a map, first figure out the main idea. The main idea is often stated in the map's title. The main idea of this map is the temperature and weather on January 13.

> **Strategy** Look at the map. Ask yourself: What is the main idea?
>
> 1. Look at the title or caption. It tells you the main idea of the map.
>
> 2. Look at the whole map. Ask yourself: What does this map show?

Many maps have clues to help you find information:

- The long arm of the compass rose always points north. Look near Miami, Florida, for the compass rose.
- The scale shows how to measure distance. Look below Houston for the scale. For example, the distance between Dallas and Houston is about 250 miles.
- The map key explains how things are shown. Look on the lower left for the key. The map key symbols show different types of weather.

 Strategy Look at the map. Ask yourself: What detailed information does this map show?

1. Look for the compass rose. It points north.

2. Look for the scale. It helps you measure distance between places on the map.

3. Look for the key. It explains how things are shown.

Exercise: Use the weather map to answer these questions.

1. Is it raining or snowing in Memphis?

   _____

2. Which city is almost directly south of Dallas?

   _____

3. Which is closer to New York—Boston or Raleigh?

   _____

4. Which is warmer—Oklahoma City or Kansas City?

   _____

5. What is the temperature in Atlanta?

   _____

Check your answers on page 163.

Lesson 11 Pollution and Recycling

Each day, for every person in the United States, 4.4 pounds of garbage are thrown out. We throw out paper, plastics, metals, glass, food, and many other kinds of trash. In one year, we throw away more than 200 million tons (over 400,000,000,000 pounds!) of garbage. Where does it all go?

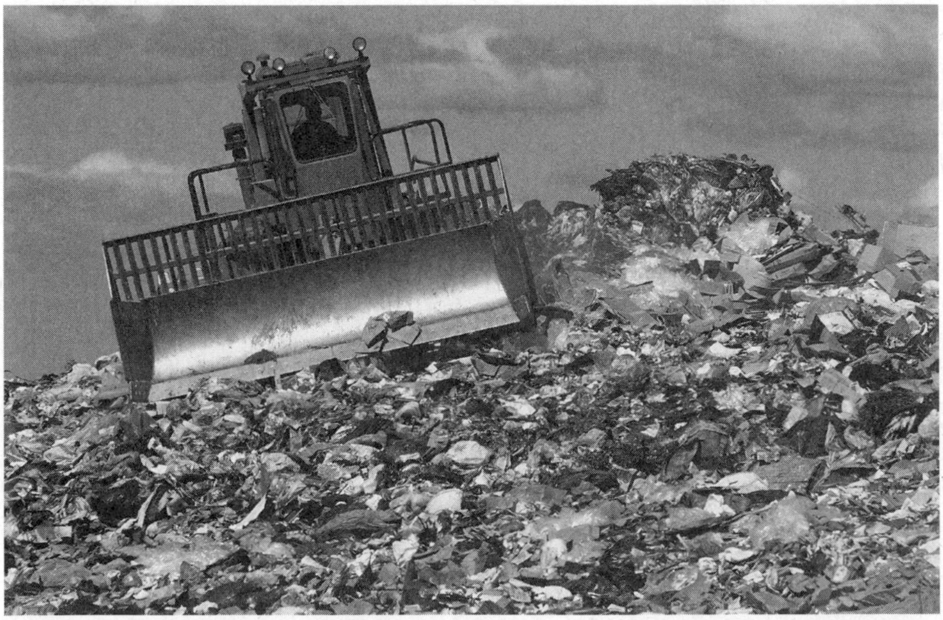

pollute
to give off harmful substances

landfill
a place where garbage is buried

incinerator
container where garbage is burned

Garbage used to be dumped on open land or in the ocean. This has been stopped because dumping **pollutes** the land and water. The pollution then harms people and animals.

Today, more than half of the garbage in the United States goes to **landfills**. Modern landfills are designed to hold in garbage to help prevent pollution. There are more than 2,000 landfills in the United States. The amount of space available for landfills is decreasing. Only 29 states say that they have enough landfill space to last beyond 2011.

About 15 percent of our garbage is burned. Today, garbage is usually burned in **incinerators**. Modern incinerators give off less air pollution than incinerators of the past.

A landfill

Recycling helps decrease the amount of garbage that is dumped or burned. About 30 percent of our garbage is recycled. In almost 10,000 communities, people separate items to be recycled from regular garbage. Glass, plastic, metal, and newspapers usually can be recycled. To help people recycle, some communities pick up these materials at people's homes. Some communities even have a separate collection for **yard waste**. To make sure that people separate items to be recycled from their garbage, some cities like Seattle charge customers for the garbage they collect. Items to be recycled are collected free of charge.

Material to be recycled is brought to a recycling center. In communities that do not pick up materials for recycling, people can bring certain items to recycling centers. The material we recycle is made into many new products. Some new products are very different from the original.

recycling
reusing something instead of just dumping it

yard waste
clippings from lawns, raked leaves, and other plant trimmings

▶ GED Tip

Some GED Science Test questions ask you to find facts in a passage. Since some facts contain numbers, pay close attention to numbers in science passages.

This plant in Rhode Island recycles tons of aluminum, paper, glass, and plastic.

fiber
thin threads that make up fabric and other materials

insulation
a material used to hold in heat

compost
a mixture of plant waste used to make soil richer

Many products made from recycled material have this symbol.

Where recycled materials go:

- Plastic soft-drink bottles are melted and made into pipes and bottles. The plastic can also be made into **fiber, insulation**, yarn, and playground equipment.
- Plastic milk containers are chopped, melted, and used to make toys, pails, and bottles for shampoo and detergent.
- Glass is sorted by color and then melted. It is formed into new bottles and jars.
- Aluminum cans are melted and made into new cans, cooking pots, lawn chairs, and siding for houses.
- Steel cans are melted. The steel is used for cans, tools, and some auto parts.
- Newspaper and other paper is mixed with water to make pulp. The pulp is used to make paper, cardboard, egg cartons, and building materials.
- Yard waste is used to make **compost**.

Recycling also includes buying things that can be recycled. Articles made from glass and paper can be recycled, but some plastic items cannot be recycled. Recycling also means buying things made out of recycled materials. Cereal boxes, paper egg cartons, aluminum cans, and glass jars are usually made with recycled materials. When you buy things made from recycled materials, you keep the recycling going.

Recycling History

In the past, people repaired and recycled things regularly. For example, during World Wars I and II, people saved metals and cooking grease to make war supplies. As time passed, people got into the habit of using things once and then throwing them out. Today, even when something could be reused, someone may throw it out instead of recycling it.

Practice

Vocabulary ■ Write the word that best completes each sentence.

1. _____ is reusing an item instead of throwing it away.

2. Yard waste can be used to make _____.

3. A _____ is a place where garbage is buried.

Finding Facts ■ Circle the number of the correct answer.

4. Which of the following is an example of pollution?

   (1) reusing newspapers
   (2) dumping garbage in the ocean
   (3) having a separate collection for yard waste
   (4) using a landfill

Reading a Map ■ Circle the number of the correct answer.

Number of Incinerators and Landfills in the United States by Region

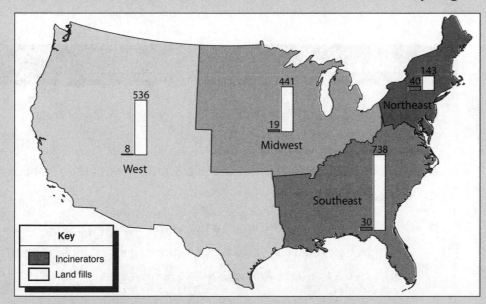

5. Which region has the most landfills?

   (1) Midwest
   (2) Northeast
   (3) Southeast
   (4) West

Check your answers on page 164.

At 4:31 A.M. on January 17, 1994, people in Northridge, California, were awakened by the loud rumbling and shaking of an earthquake. For 40 seconds, the Los Angeles suburb shook. All over the city, buildings swayed and things came crashing down. Oil and gas lines broke, causing explosions. Tens of thousands of homes lost water or electricity. Major highways collapsed. Those 40 seconds caused over 15 billion dollars worth of damage. All together, the Northridge earthquake killed 55 people and injured 4,000.

The Northridge earthquake started in the rocks directly below the city of Northridge. The Northridge earthquake caused so much damage because there were many people and buildings close together. The Northridge earthquake was medium sized. It was not "The Big One." The Big One is a major earthquake that scientists predict will occur in California sometime in the next 30 years.

Earthquake Facts

- There are several million earthquakes every year. Many of them are small or occur where few people live.

- The most destructive earthquake in U.S. history was in San Francisco on April 18, 1906.

- One of the deadliest earthquakes in the world occurred on December 26, 2004. The earthquake happened in the Indian Ocean and caused a **tsunami**. Almost 300,000 people died.

tsunami
a large sea wave made by an underwater earthquake or volcano

Earthquakes are caused by movements of Earth's **crust**. Earth's crust is not one piece. It is made up of about a dozen large **plates**, or sections, and many smaller plates. These plates are not locked together. Instead, the plates float on the hot, melted **mantle** beneath them. At the center of Earth is its **core**.

The different plates that form Earth's crust don't all float around in the same way. Each plate has its own speed and direction. The Pacific plate under the Pacific Ocean is moving to the northwest. The North American plate is moving west. As the Pacific and North American plates slide past each other, they sometimes get "stuck." When this happens, the pressure between the plates builds up until there is an earthquake. Then the plates loosen and continue on their way.

crust
the outer layer of Earth

plate
a section of Earth's crust that floats on the mantle

mantle
the part of Earth between the crust and the core that is liquid and very hot

core
the center of Earth

Movement of the Pacific and North American Plates

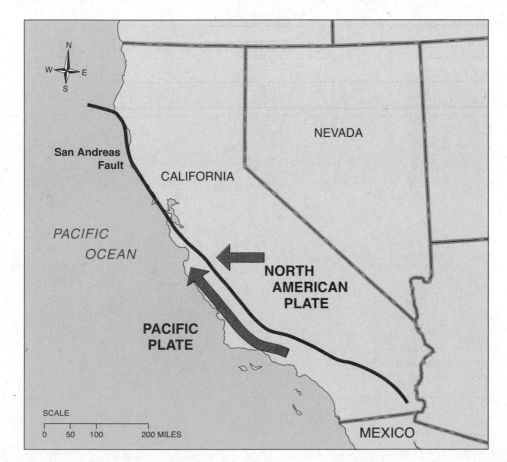

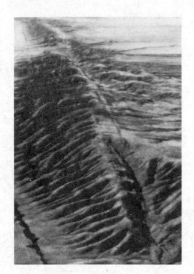

*The San Andreas
fault can be seen for
hundreds of miles.*

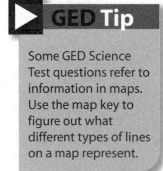
Predicting where an earthquake will occur is not very difficult. In California, you can actually see where the two plates meet along the San Andreas **fault**. The fault runs more than 800 miles, from Mexico to Cape Mendocino. Earthquakes in California usually occur on or near the San Andreas fault. The Northridge earthquake of 1994 was west of the San Andreas.

Predicting when an earthquake will occur is more difficult. Along the San Andreas fault, earthquakes happen when the plates very suddenly start sliding. Scientists have located areas along the fault that have been stuck for more than a century. Scientists know pressure is building in these places. However, they don't know how much pressure there is or when it will be released. The Big One could happen in any of these places at any time.

Major Faults in California

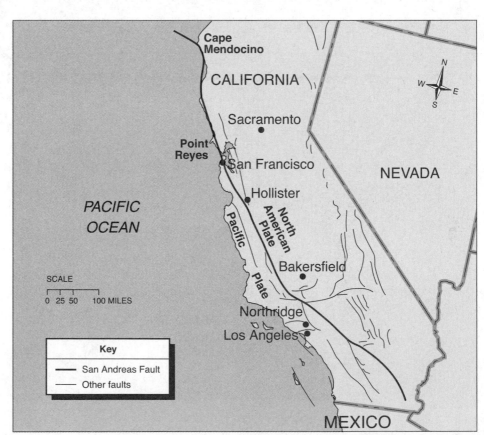

Practice

Vocabulary ■ Write the word that best completes each sentence.

1. The part of Earth between the crust and the core is the
   _____.

2. A crack in rock along which plate movement occurs is a
   _____.

3. Movements of Earth's crust can cause an
   _____.

earthquake

fault

crust

mantle

Finding the Main Idea ■ Circle the number of the correct answer.

4. What is the main idea of the first paragraph on page 72?

   (1) Earthquakes in California usually occur near the
       San Andreas fault.
   (2) Predicting where an earthquake will occur is not very difficult.
   (3) The Northridge earthquake was west of the San Andreas fault.
   (4) Predicting when an earthquake will occur is not very difficult.

5. What is the main idea of the last paragraph on page 72?

   (1) Earthquakes happen when the plates very suddenly start
       sliding again after being stuck.
   (2) "The Big One" could happen at any time.
   (3) Predicting when an earthquake will occur is difficult.
   (4) Some areas along the fault have been stuck for more than
       a century.

Reading a Map ■ Look at the map of California on page 72 to answer the questions. Then write your answers.

6. Are most of the faults located in northern or southern California?

   _____

7. What are the names of two cities located on or near a fault?

   _____

Check your answers on page 164.

Using Context Clues

If you were reading and you came to a word you didn't understand, what would you do? You might look it up in a dictionary. However, there is a faster way to try to find its meaning. Sometimes you can look at the other words in the sentence or paragraph and figure out what the word means. In other words, you can figure out the word from its context. See if you can figure out the meaning of the word *gyring* in the following paragraph:

> If the hurricane stays over warm seas, it may grow until the eye—the calm area in the middle of the storm—is 60 miles across. The hurricane winds, in contrast, are gyring wildly around at distances up to 300 miles from the center.

The paragraph tells you that the area in the middle of the storm is calm. The winds are in contrast to that calmness. *Wildly* gives you another clue that *gyring* means "moving quickly."

 Strategy Look for word clues. Ask yourself: What do the other words in the paragraph mean?
1. Find words that show how things are the same or different.
2. Find words that describe the word that you want to know the meaning of.

Exercise 1: Look for the word *spewed* in the following paragraph. Then write what you think it means.

> In August 1883, Krakatoa, a volcano near Java, exploded. It spewed nearly five cubic miles of rock and ash violently into the air.

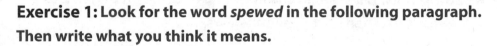

Krakatoa

Exercise 2: Look for the term *curbside recycling* in the following paragraph. Then write what you think it means.

Recently, curbside recycling programs have spread throughout the country. In many communities, people separate regular garbage and other items. Usually, glass, plastic, metal, and newspapers are packaged separately and collected for recycling. One truck may collect the regular garbage. A second truck collects the glass, metal, plastic, and paper. These items are taken to factories where they are cleaned and used again.

_____

_____

 Strategy Think it through. Ask yourself: How do the ideas in the paragraph fit together?

1. Read and reread the words or sentences around the unknown word.

2. Use what you already know to figure out the meaning of the word.

Exercise 3: Look for the word *radiosonde* in the following paragraph. Then write what you think it means.

The upper atmosphere is not an area that meteorologists can easily reach. That's why a radiosonde is so useful. Carried high above Earth by a helium-filled balloon, a radiosonde sends out radio signals. A special kind of radio back on the ground receives the signals. As a result, a meteorologist can tell what the weather is like high above the clouds.

A meteorologist

_____

_____

Check your answers on page 164.

Lesson 13   The Solar System

universe
space and everything in it

The **universe** is larger than we can imagine. What is it made of? How did it come to be? What is Earth's place in the universe?

The Solar System

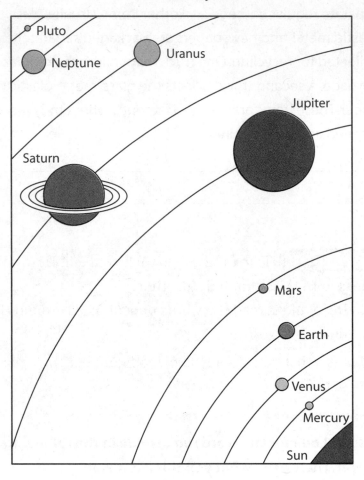

The universe probably began about 14 billion years ago with an explosion called the **Big Bang**. Something very small, hot, and dense exploded with such force that even today, gas and dust are still moving outward in all directions.

Big Bang
the explosion that may have started the universe

solar system
our sun and its nine planets

One giant spinning cloud of gas and dust became our **solar system**. About five billion years ago, the spinning cloud shrank as most of the gas was pulled into the center and became our sun. Around the sun, gas and dust clumped into nine planets including Earth, dozens of moons, and thousands of smaller objects.

Why don't Earth and the other eight planets in our solar system fly off into space? **Gravity** holds the solar system together. The more **mass** an object has, the more it pulls on other objects. More than 99 percent of the solar system's mass is in the sun. The planets in our solar system are pulled toward the sun.

Why don't the planets all fall into the sun? All objects resist change in motion. Gravity and the planets' resistance to change balance each other. Each planet is pulled in a path around the sun. This path is called the planet's **orbit**. The time it takes to complete an orbit is a planet's year. Earth's year is 365 days long.

The sun and its nine planets are all spinning. Earth spins once every 24 hours. It faces the sun during the day. It faces away from the sun during the night. Each spin, or day, is called a **rotation**. Different planets spin at different speeds. For example, Venus takes 243 days to complete one rotation.

gravity
the force that pulls any two objects together

mass
the amount of material in an object

orbit
the path of a planet around the sun

rotation
the time it takes a planet to spin one complete turn on its axis

Planet Days

Period of Rotation in Earth Days: 59

Mercury

Period of Rotation in Earth Days: 243

Venus

Period of Rotation in Earth Days: 1

Earth

Period of Rotation in Earth Days: 10 hours

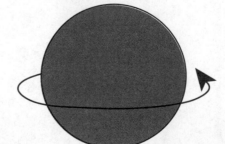

Jupiter

seasons
yearly changes in temperature caused by the tilt of a planet's axis as it moves around the sun

axis
an imaginary line from the North Pole to the South Pole

Most places on Earth have four seasons. The tilt of Earth's axis as it moves around the sun causes the seasons.

galaxy
a group of stars held together by gravity

Milky Way
the name of the galaxy where our sun and Earth are found

Goldilocks Conditions
the conditions needed for life as we know it

GED Tip

Some GED Science Test questions ask you to compare two things. Pay attention to key details that make one thing different from the other.

As Earth orbits the sun, each place on Earth goes through cycles of warmer and colder weather. These yearly changes are called **seasons**. Earth has seasons because it is tilted. Imagine a line from the North Pole to the South Pole. This line is Earth's **axis**. Earth's axis tilts toward the sun during the warm part of the year and away from the sun during the cold part.

The Reason for Seasons

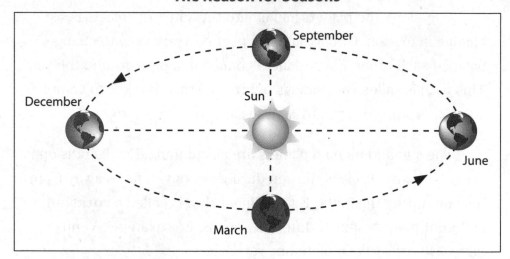

Our sun is just one of about 200 billion stars in our **galaxy**, the **Milky Way**. The Milky Way is just one group of stars among at least 100 billion galaxies. What makes Earth so special?

Earth may be the only place in the universe that supports life as we know it. To support life, the air has to have the right mix of gases. Water must exist as a liquid, not just as ice or gas. The temperature must not be too hot or too cold. This sounds like the children's story of *Goldilocks and the Three Bears*. That's why we say that a planet with water and the right air and temperature meets the **Goldilocks Conditions**.

Practice

Vocabulary ■ Write the word or words that best complete each sentence.

1. The universe probably began with an explosion called the

   _____.

2. A planet's path around the sun is its

   _____.

3. The galaxy we live in is called the

   _____.

> **Big Bang**
> **Goldilocks Conditions**
> **Milky Way**
> **orbit**

Finding Facts ■ Circle the number of the correct answer.

4. Based on information in the first paragraph on page 77, what is the biggest object in our solar system?

   (1) Earth
   (2) moon
   (3) sun
   (4) planets

5. Based on information in the second paragraph on page 77, one Earth orbit around the sun takes

   (1) 1 day.
   (2) 7 days.
   (3) 30 days.
   (4) 365 days.

6. Based on the graphic on page 77, which of the following planets has the longest day?

   (1) Earth
   (2) Venus
   (3) Jupiter
   (4) Mercury

Check your answers on page 164.

constellation
an imagined picture
made of stars

For thousands of years, people have imagined that pictures tell stories in the night sky. The pictures are like giant connect-the-dots drawings, but the dots are stars, or balls of hot, glowing gas. We call these star pictures **constellations**.

The sky at night really does tell a story. It tells the story of the universe since time began. Some stars are so far away that it takes many years for their light to reach our eyes. Looking into the night sky is like looking back in time.

The Night Sky

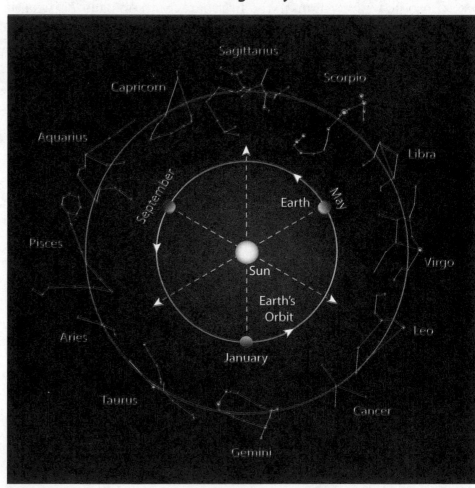

People divided the sky into different pictures in order to make a single star easier to find.

You can use the diagram above to figure out which constellations are visible during certain months. For example, if you follow the arrow painting from Earth during May, you see that Aries is visible at that time.

A star begins when gravity pulls enough gas and dust close together. When the gas and dust become dense enough, the star gives out a huge amount of heat and light. After billions of years, the star uses up its energy and disappears. Some stars explode. Some just fade away.

The stars in a constellation may all look as if they are the same distance from Earth, but they are not. A bright star that is far away from Earth may look the same to us as a dim star that is closer to Earth. Almost all the stars that we see in the night sky are in our own galaxy, the Milky Way.

Stars are always moving, so constellations change slowly over many years. A million years ago, the **Big Dipper** looked like a spear. Today it looks like a cup with a handle. Nobody knows what it will look like a million years from now.

Big Dipper
a constellation made up of seven stars

The Changing Big Dipper

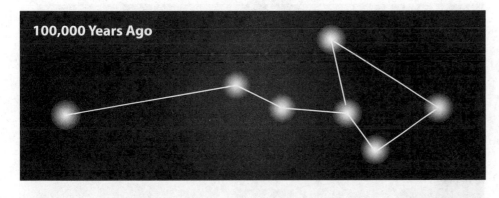

100,000 Years Ago

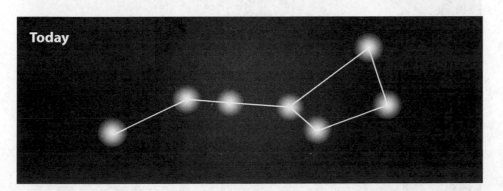

Today

Constellations change very slowly over many years.

Earth takes 365 days and nights to travel around the sun. Because Earth is moving, we see different parts of the night sky at different times of the year. For each of the 12 months, a different constellation is in the middle of the sky. This ring of 12 constellations is called the **zodiac**.

Throughout history, we have used constellations to help us. Travelers, especially sailors far from land, use the stars as guides. The arrival of the same constellation year after year also reminds farmers that it is time to plant or harvest the crops.

zodiac
the 12 monthly constellations that follow each other across the middle of the night sky

The Southern Cross

Some stars can be seen only from the north or the south of Earth. The Southern Cross, the smallest constellation, can only be seen in the south. The long arm of the cross points south. The Southern Cross has helped guide sailors for hundreds of years.

The Southern Cross

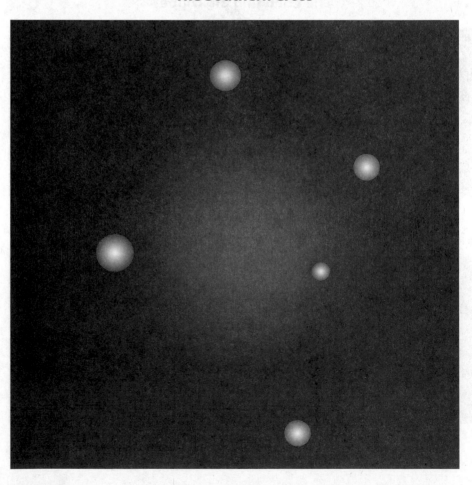

GED Tip

When you read a GED Test question, try to answer it before you read the choices. If one of the answers is similar to your answer, it is probably correct.

Practice

Vocabulary ■ **Write the word or words that best complete each sentence.**

1. Star pictures are called _____.

2. The ring of 12 star pictures in the middle of the sky over the course of a year is the _____.

3. The smallest constellation is the _____.

constellations
zodiac
Southern Cross
Big Dipper

Using Context Clues ■ **Circle the number of the correct answer.**

4. In the first paragraph on page 81, *dense* probably means

    (1) close together.

    (2) hard.

    (3) old.

    (4) cold.

Reading a Map ■ **Circle the number of the correct answer.**

5. The map on page 80 shows the night sky. During which month is Sagittarius directly overhead Earth?

    (1) May

    (2) September

    (3) January

    (4) July

Check your answers on page 165.

GED Test-Taking Strategy

Previewing Graphics

On the GED Science Test, you answer questions based on graphics such as diagrams, graphs, and maps.

One strategy for passing the GED Test is to preview the graphics before you read the questions. This will help you understand what the question is asking.

 Strategy Try this strategy on the example below. Use these steps.

Step 1 Look at the graphic. Read the title and labels.

Step 2 Read the question.

Step 3 Look at the graphic again. Look for what the question asked you to find out.

Step 4 Answer the question.

Inside the Sun

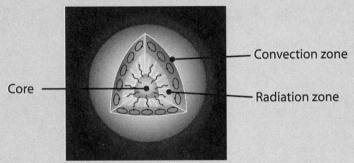

According to the title, what is the graphic about?

(1) the distance between the sun and planets

(2) the parts of the sun

(3) the size of the sun

(4) the heat that the sun gives off

In Step 1 you looked at the graphic and read the title and the labels. In Step 2 you read the question. It asked you to use the title to understand what the graphic was about. In Step 3 you looked at the graphic again while thinking about the question. The correct answer is (2). Choices (1), (3), and (4) are not related to the title.

Practice the strategy. Use the steps you learned. Circle the number of the correct answer.

Where and When to Expect Tornados

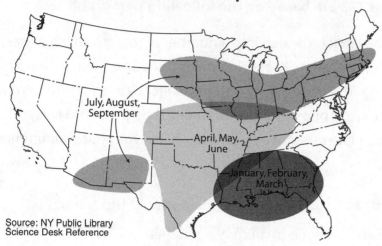

July, August, September

April, May, June

January, February, March

Source: NY Public Library
Science Desk Reference

1. What is the main idea of the graphic above?

  (1)   Tornadoes occur in winter in the southeast.

  (2)   Many states have tornadoes in summer.

  (3)   Tornado season is different in different areas.

  (4)   Many states have a tornado season.

Inside Earth

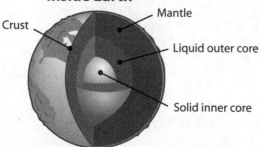

Crust

Mantle

Liquid outer core

Solid inner core

2. Moving from the outside in, what is the correct order of Earth's layers?

  (1)   crust, mantle, outer core, inner core

  (2)   mantle, crust, outer core, inner core

  (3)   inner core, outer core, crust, mantle

  (4)   mantle, inner core, outer core, crust

Check your answers on page 165.

Read each paragraph and question carefully. Circle the number of the correct answer.

Questions 1–3 are based on the following paragraph.

Earth's crust is divided into a number of regions called plates. Each plate is between 20 and 150 miles thick. They float on a bed of molten rock. The plates move only a few inches each year. When two plates going in opposite directions meet, they crash into each other. Where one plate rides up over another plate, volcanoes sometimes form.

1. Which of the following is the main idea of this paragraph?

   (1) Earth is made up of moving plates.

   (2) Earth's plates move a few inches per year.

   (3) When two plates meet, they crash into each other.

   (4) Volcanoes are found where one plate rides up over another plate.

2. Based on the information in the paragraph, the word *molten* probably means

   (1) solid.

   (2) frozen.

   (3) melted.

   (4) hard.

3. You are likely to find volcanoes

   (1) where plates are stable.

   (2) in the middle of a plate.

   (3) where one plate rides up over another plate.

   (4) where two plates are moving in opposite directions.

Where Earthquakes Are Likely to Cause Damage

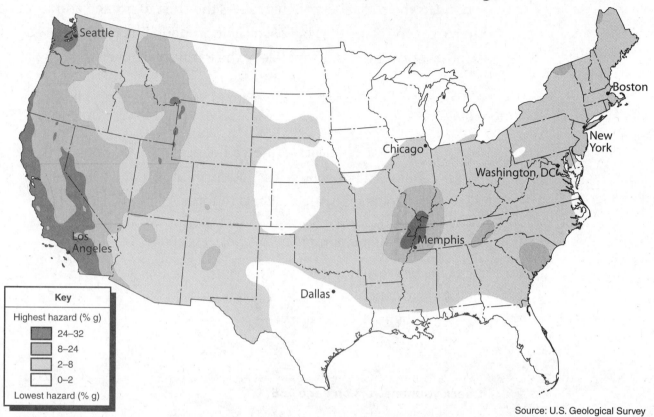

Key

Highest hazard (% g)

- 24–32
- 8–24
- 2–8
- 0–2

Lowest hazard (% g)

Source: U.S. Geological Survey

4. The main point of the map is to show

   (1) where the next earthquake will occur.
   (2) that most earthquake damage occurs near big cities.
   (3) that most of the U.S. is safe from earthquakes.
   (4) which places are likely to be damaged by earthquakes.

5. Which of the following cities is least likely to be damaged by earthquakes?

   (1) Boston
   (2) Chicago
   (3) Dallas
   (4) Seattle

Questions 6–7 are based on the following paragraph.

A hot spot is a place on the earth that is like a blowtorch. The hot spot heats the crust above it. This causes the crust to expand and bubble up. The result is a kind of volcanic eruption. The Hawaiian Islands were formed in the middle of the Pacific plate over a hot spot.

6. Which of the following is the main idea of this paragraph?

   (1) A hot spot is like a blowtorch.
   (2) The Hawaiian Islands sit on a hot spot.
   (3) There is a hot spot in the middle of the Pacific plate.
   (4) Hot spots may cause islands to form.

7. Based on the paragraph, a *hot spot* is probably

   (1) a blowtorch.
   (2) a volcano.
   (3) a hot area inside Earth.
   (4) an island in the ocean.

Check your answers on page 165.

Unit 2 Skill Check-Up Chart

Check your answers. In the first column, circle the numbers of any questions that you missed. Then look across the rows to see the skills you need to review and the pages where you can find each skill.

Question	Skill	Page
3	Finding Facts	Unit 1, pages 16–17
1, 6	Understanding the Main Idea	Unit 1, pages 40–41
4, 5	Reading a Map	pages 64–65
2, 7	Using Context Clues	pages 74–75

Unit 3 Chemistry

In this unit you will learn about

- atoms and elements
- states of matter
- mixtures and solutions
- acid rain

molecule

acid

element

solvent

Chemistry is the study of what everything is made of. Chemistry is part of what you can see, touch, feel, and smell.

What is something in your home that contains chemicals?

____

____

Chemistry is also the study of how materials change. Every time you make ice cubes or bake a cake, you experience chemistry.

What is one way that you can change water to another form?

____

____

matter
a substance that occupies space and can be seen, sensed, or measured

chemistry
the study of matter

element
a substance that can't be broken down into simpler substances

Stub your toe on a wooden chair, and you know just how hard wood can be! Dip your toe in a tub of warm water, and your toe will feel warm and wet. Wood and water are different, but they are both **matter**. Your toe, this book, rain, and even air all are made of matter.

Matter is everywhere. Scientists try to understand it by studying **chemistry**. Did you ever wonder why water drops form on the outside of a glass of cold water? Do you know why wood and water feel so different from each other? Chemists answer these questions by studying matter.

Matter is made up of many different substances called **elements**. You may already be familiar with some elements. Carbon is a common element. You find carbon in diamonds and in the "lead" in your pencil. Carbon is also found in all living things, including trees, animals, and people. Gold and silver are elements that have been used for centuries to make jewelry and coins. Oxygen is an element found in air, water, crystals, glass, and sand.

Air, water, flowers, wood, and sand are all forms of matter.

The smallest parts of an element are called **atoms**. Scientists believe the structure of an atom looks like our solar system. The solar system has a sun in the middle and planets that move around, or orbit, it. An atom has a **nucleus** in the middle and particles, or small parts, that move around it. The nucleus contains two kinds of particles, **protons** and **neutrons**. **Electrons** orbit the nucleus in a kind of cloud. An electron has a very small **mass**. Protons and neutrons also have small masses, but they each have about 1,800 times the mass of an electron.

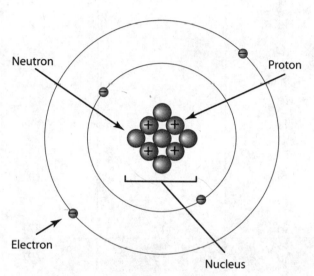

Atoms are made up of protons, neutrons, and electrons.

Electrons move around the nucleus at high speeds. They are found at different distances from the nucleus. The farther an electron is from the nucleus, the more energy it has.

All atoms of a certain element have the same number of protons and electrons. For example, all carbon atoms contain 6 protons and 6 electrons. All gold atoms have 79 protons and 79 electrons. No two elements have the same number of protons. Therefore, this number of protons, which is called an element's **atomic number**, is different for every element.

atom
the smallest part of an element

nucleus
an atom's center that contains protons and neutrons

proton
a positively charged particle in an atom's nucleus

neutron
an uncharged particle in an atom's nucleus

electron
a negatively charged particle in an atom

mass
the amount of material in an object

atomic number
the number of protons in an atom

The table below lists some common elements. Notice that these elements contain different numbers of protons and electrons. Different kinds of matter have different features. Their **atomic structure** is the reason for this. The atoms that make up different kinds of matter are not all the same. An atom in an aluminum can and an atom in a gold ring have different numbers of particles. Therefore, the atoms also have different atomic numbers.

Table of Elements			
Element	Symbol	Atomic Number (Number of Protons)	Number of Electrons
Hydrogen	H	1	1
Helium	He	2	2
Carbon	C	6	6
Oxygen	O	8	8
Neon	Ne	10	10
Sodium	Na	11	11
Aluminum	Al	13	13
Silicon	Si	14	14
Chlorine	Cl	17	17
Silver	Ag	47	47
Gold	Au	79	79

Notice that each element in the table has a **symbol** made of one or two letters. These symbols make it quicker to write the elements.

Everything around you is made of atoms. Some materials are made up of only one kind of atom. Because pure gold is made of only gold atoms, gold is an element. Water is made of both oxygen and hydrogen atoms, so it is a **compound**. You are probably familiar with many other compounds. Table salt is a compound made up of the elements sodium and chlorine. It is called sodium chloride. The table below lists some common compounds.

Compound	Elements
Salt	Sodium, Chlorine
Sugar	Carbon, Hydrogen, Oxygen
Glass	Silicon, Oxygen

GED Tip

Some GED Science Test questions ask you to find information in a table. Use the table title and headings to help you find the information.

Practice

Vocabulary ■ **Choose the word or words that best complete each sentence.**

1. The _____ carbon is a common substance on our planet.

2. Sodium hydroxide is a _____ made up of the elements sodium, hydrogen, and oxygen.

3. Each element has a different _____ number.

4. Atoms contain _____ with a positive charge and _____ with a negative charge.

> atomic
> compound
> electrons
> element
> neutrons
> protons

Finding Facts ■ **Circle the number of the correct answer.**

5. A compound is made up of

   (1) atoms that all are alike.
   (2) elements that all are alike.
   (3) two or more elements.
   (4) two or more atomic numbers.

6. The nucleus of an atom is made up of

   (1) electrons and protons.
   (2) protons and neutrons.
   (3) neutrons and electrons.
   (4) electrons, protons, and neutrons.

7. Based on the last paragraph on page 92, you can conclude that water is

   (1) an atom.
   (2) an element.
   (3) a compound.
   (4) a proton.

Check your answers on page 166.

state
a form of matter: solid, liquid, or gas

solid
state of matter that has a volume and shape that can't change

liquid
state of matter that has a volume that can't change but a shape that can change

gas
state of matter that has a volume and shape that can change

H_2O
the chemical formula for water

molecule
two or more atoms connected together

vibrate
shake rapidly

If you take an ice cube out of the freezer and leave it on the kitchen table, it melts. It changes from a solid to a liquid. This is a change of **state**. If you put the water in a pan and heat it, the water will change state again. It will become water vapor, a gas. Matter can change state when there is a change in temperature. The three states of matter are **solid**, **liquid**, and **gas**.

Ice, water, and water vapor are the same compound but in different states. They are all **H_2O**. Two atoms of hydrogen (H_2) and one atom of oxygen (O) become connected to form a **molecule** of water. Ice is H_2O, liquid water is H_2O, and water vapor is H_2O.

Three states of matter: solid, liquid, and gas

Ice   Water   Water Vapor

What happens when H_2O changes state? When water is in a solid state (ice), its molecules are lined up in an orderly way. The molecules of solid H_2O **vibrate** in their places. Ice must be heated to change its state. The warmer the ice becomes, the faster the molecules vibrate. At some point during the heating, the molecules shake so fast that they don't stay in their places anymore. We call this movement "melting." H_2O has changed from a solid state (ice) to a liquid state (water). It is still the same substance. Each molecule has exactly the same atoms in it.

What happens as water in a pan is heated and becomes water vapor? As the water gets hotter, the molecules move faster and faster. They bounce off each other and off the sides of the pan. We say the water is "boiling." The molecules near the surface of the water begin to bounce out into the air. They mix with the atoms in the air. At some point there are no water molecules left in the pan. All of the H_2O molecules still exist, but they have moved into the air. The H_2O has changed from a liquid (water) to a gas (water vapor).

Two ways to measure temperature

Fahrenheit (F) and Celsius (C) are two different scales that measure temperature. Most people in the United States use Fahrenheit to measure temperature. However, most people in the rest of the world use the Celsius scale. Scientists in the United States often use the Celsius scale in their work.

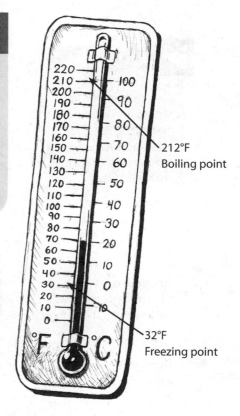

212°F
Boiling point

32°F
Freezing point

This thermometer shows temperature degrees in both Fahrenheit and Celsius.

We see water change state, from solid to liquid to gas, in our daily lives. Water freezes at 32°F (0°C) and boils at 212°F (100°C). A normal kitchen freezer and stove will give us these temperatures. Most other substances do not change state at these temperatures. For example, you can't melt an iron frying pan on the stove.

We think of most metals as solids because they are solids at the temperatures we live in. However, all metals melt if they're brought to a high enough temperature. Tin melts at 449°F. Iron must be heated to 2,795°F to melt. If a metal is heated enough, a liquid metal will boil away into a gas. Lead boils at a temperature of 3,180°F.

Solid   Melting Point →   **Liquid**   Boiling Point →   **Gas**

Freezing Point

When the particles in a substance are heated, they gain energy and move faster.

▶ **GED Tip**

Pay attention to the arrows and labels in a diagram. They often explain a process, such as how a substance changes state.

tungsten
a hard metal that has a very high melting point

filament
a thin, flexible wire

mercury
a poisonous silver-colored metal that has a low melting point

If you heat a solid, it will melt and then boil. This process can be reversed. Gases become liquids and then solids as they get colder. We think of oxygen as a gas because that is its state at room temperature. However, if we cool oxygen to –297°F, it becomes a pale blue liquid. If we cool it even further, it freezes into a solid at –361°F. If solid oxygen is heated, it will begin to melt at –361°F. The temperature at which a substance melts or freezes is the same. For oxygen, this temperature is –361°F.

The fact that different elements have different melting (and freezing) points is very important in our daily lives. The metal with the highest melting point of all is **tungsten**. It doesn't melt until it reaches a temperature of 6,192°F. This is why tungsten is used to make **filaments** in light bulbs.

Mercury, on the other hand, is a metal that is a liquid at room temperature. Mercury is sometimes used in thermometers to show the temperature. Mercury freezes at –38°F. If the temperature drops lower than that, the thermometer won't work. Since mercury is poisonous and only can be used to temperatures of –38°F, alcohol thermometers are more useful. Alcohol doesn't freeze until the temperature drops to –202°F.

No matter what the substance is, all molecules stop moving at –459°F. We call this point absolute zero. Matter cannot be colder than this temperature.

Practice

Vocabulary ■ Choose the word that best completes each sentence.

1. The metal with the highest melting point is
   _____.

2. At 5,000°F, this metal is in a solid _____.

3. It melts to form a _____ at 6,192°F.

gas

liquid

state

tungsten

Using Context Clues ■ Circle the number of the correct answer.

4. In the last paragraph on page 94, *vibrate* means to

   (1) line up in an orderly pattern.
   (2) break down.
   (3) move back and forth.
   (4) melt.

5. In the second paragraph on page 96, a *filament* is

   (1) a metal with a high melting point.
   (2) a metal with a low freezing point.
   (3) a liquid in a light bulb.
   (4) a tiny wire that lights up.

Finding Facts ■ Circle the number of the correct answer.

6. When H_2O is in its solid state, its molecules are

   (1) different from water.
   (2) different from water vapor.
   (3) lined up in an orderly way.
   (4) not moving.

7. Some states of matter are

   (1) solid.
   (2) liquid.
   (3) gas.
   (4) all of the above

Check your answers on page 166.

GED Skill Strategy

Understanding Cause and Effect

A *cause* makes something happen. An *effect* happens as a result of a cause. The sentence below shows cause and effect.

The water boiled *because* I heated it.

When you read, look for words that tell you that one thing is causing another thing. For example, the word *because* in the sentence above tells you something caused the water to boil. The cause is heat. The effect is boiling.

 Strategy Look for key words and phrases. Ask youself: Which words show cause and effect?

1. Pay attention to these words and phrases: *so*, *for*, *because*, *cause*, *effect*, *make*, *as a result*, *for this reason*, *consequently*, *therefore*, *when*, and *if*.

2. Use what you know to figure out what the cause is and what the effect is.

Exercise 1: Read the paragraph. Then answer the questions.

Bakers add live yeast and sugar to bread dough. The yeast break down the sugar, causing alcohol and carbon dioxide to form. The carbon dioxide gas makes the bread dough rise. The yeast make more and more carbon dioxide. The effect is expanding dough. The dough is then baked. As a result of the heat, the yeast die and the alcohol evaporates.

The word *causing* in the second sentence tells you that a result follows. Yeast breaking down sugar is the cause. The effect is that alcohol and carbon dioxide form.

1. What causes the bread dough to rise? _____

2. What word helped you find the cause? _____

3. What causes the alcohol in the bread dough to evaporate?

_____

> **Strategy** Figure out from the context what causes something to happen. Ask youself: How are the actions related?

1. Read and reread the paragraph.
2. Look for clues such as the order in which actions take place.
3. Use what you know to figure out what the cause is and what the effect is.

Exercise 2: Read the paragraphs. Then answer the questions.

Why do many cities spread salt on the roads during winter weather to make them safer to drive on? When the temperature falls below 32°F, water on roads freezes, and the roads become slippery.

When one substance (salt) is mixed into another substance (water) until the first substance disappears, the boiling and freezing points of that combined substance are changed. Adding salt to icy roads lowers the freezing point of the water. Therefore, the water remains in a liquid state, and the road is not so slippery. Spreading salt on the roads can keep the water in a liquid state down to about 14°F. When temperatures fall below 14°F, salt doesn't mix with water as easily, so salting the roads does not work very well.

1. What is the effect of adding salt to water?

   _____

2. What happens to water on roads when salt is added?

   _____

3. Why does salting roads not work well below 14°F?

   _____

Check your answers on page 166.

When matter is pure, it is not mixed with other types of matter. For example, pure gold has only one type of atom: gold atoms. Pure table salt contains only one compound: sodium chloride. It is not mixed with other types of compounds.

Most things in nature are mixed. When you go to the beach, the water and sand are not pure. Ocean water contains salt and other minerals. Sand contains pieces of shell, rock, and seaweed all mixed together.

mixture
a combination of two or more substances

The sand on the beach is a **mixture**. You can see the parts of the mixture and then separate them from one another. Granola, a bag of marbles, and a salad are also mixtures. Granola can be separated into oats, raisins, and nuts. A bag of marbles can be separated by size or color.

A salad is a mixture. It can be separated into its parts: lettuce, tomatoes, and carrots.

Sometimes it is not easy to see how a mixture can be separated. Muddy water is another example of a mixture. Although the mud and the water seem to be completely mixed together, you can separate the dirt from the water. You can let the dirt settle out of the water. Or you can remove the dirt from the water by using a filter.

Ocean water is a special type of mixture called a **solution**. In a solution the parts are completely mixed, and the different parts cannot be seen. They cannot be easily separated. You cannot see the salt and other minerals in ocean water.

You can see how a solution is made when you mix sugar with water. When sugar is mixed with water, the sugar mixes evenly throughout the water and disappears. The sugar **dissolves** in the water.

In a sugar water solution, sugar is the **solute**. It is the substance that disappears when the solution is made. The substance that dissolves the solute is the **solvent**. The solvent is present in a larger amount than the solute. Water is the solvent in a sugar water solution.

solution
a mixture in which the substances are completely mixed and the different parts cannot be seen

dissolve
when one substance disappears into another to make a solution

solute
the substance that disappears in another substance (the solvent)

solvent
a substance, such as water, that dissolves another substance (the solute)

1. *A solution can be made with sugar and water.*

2. *Water is the solvent and sugar is the solute.*

3. *The sugar dissolves in the water.*

4. *The sugar cannot be seen in the sugar water solution.*

Mixtures and solutions can be made with solids, liquids, and gases. Rubbing alcohol is a mixture of two liquids, alcohol and water. The air we breathe is a mixture of gases. Gold jewelry is made from two solids, gold and silver. Sugar water is a solution made with a solid, sugar, and a liquid, water.

▶ **GED Tip**

Some GED Science Test questions refer to tables. When you look at a table, read the words at the top of the table first.

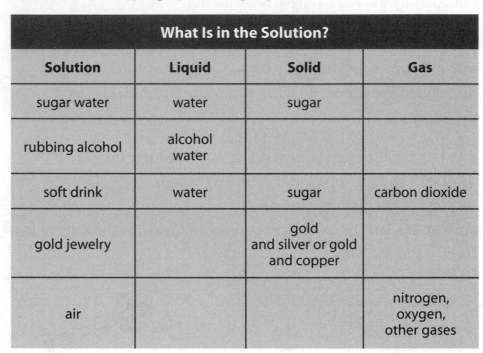

What Is in the Solution?			
Solution	**Liquid**	**Solid**	**Gas**
sugar water	water	sugar	
rubbing alcohol	alcohol water		
soft drink	water	sugar	carbon dioxide
gold jewelry		gold and silver or gold and copper	
air			nitrogen, oxygen, other gases

The final state of a solution is the same as the state of the solvent, the substance that is present in the greatest amount. For example, when you mix sugar in water, the solution is a liquid. When carbon dioxide is mixed with water, the carbonated water is also a liquid.

properties
qualities belonging to an individual or a thing

The **properties** of solutions are different from the properties of pure substances. The change in the properties of solutions can be useful. A solution of antifreeze and water keeps the water in the radiator of a car from freezing on a cold day. The temperature at which the antifreeze solution freezes is lower than the temperature at which pure water freezes.

In its pure form, gold is too soft to make into jewelry. Melted gold and silver can be mixed together. They cool into a solid solution. Mixing gold with silver makes gold harder.

Practice

Vocabulary ■ **Write the word that best completes each sentence.**

1. A combination of two or more substances is called a
   _____.

2. A mixture in which the substances are completely
   mixed and the different parts cannot be seen is called a
   _____.

3. When one substance disappears into another, it
   _____.

dissolves

mixture

solution

solvent

Finding Facts ■ **Circle the number of the correct answer.**

4. You can see and easily separate the parts of

   (1) a solution.
   (2) air.
   (3) a mixture.
   (4) a solute.

5. In a sugar and water solution, what is the solute and what is
   the solvent?

   (1) Sugar is the solute, and water is the solvent.
   (2) Water is the solute, and sugar is the solvent.
   (3) Sugar is the solute, and the solution is the solvent.
   (4) Water is the solvent, and there is no solute.

Cause and Effect ■ **Circle the number of the correct answer.**

6. Pure gold is too soft to be made into jewelry. What causes
   gold to be hard enough to make into jewelry?

   (1) melting it
   (2) mixing it with silver
   (3) dissolving it in water
   (4) mixing it with antifreeze

Check your answers on page 166.

GED Skill Strategy

Reading Tables and Bar Graphs

Tables organize information by putting it in columns, which are vertical, or up and down, and rows, which are horizontal, or side to side. At the top of each column is a column head that explains the information in that column.

To read information in tables, first find out how they are organized. Then you will be able to find the facts you need to know.

> **Strategy** Become familiar with the way the table is organized. Then find the facts you need. Ask yourself: What kind of information does each column and row show?
>
> 1. Read the column heads. Notice what each row represents.
>
> 2. Find a specific fact in the table by reading down a column and across a row.

Exercise 1: Use the table below to answer the questions.

Some Common Acids		
Name	**Chemical Formula**	**Where Found**
Sulfuric acid	H_2SO_4	in acid rain
Acetic acid	CH_3COOH	in vinegar
Hydrochloric acid	HCl	in the stomach
Citric acid	$C_6H_8O_7$	in citrus fruits

1. List the column heads shown in the table.

   _____

2. What is the chemical formula for citric acid? _____

3. Which acid is found in the human body?

   _____

Bar graphs organize information by using two axes, or lines, one horizontal and one vertical. Bar graphs combine words and numbers to explain information. The length of each bar in a bar graph stands for an amount.

 Strategy Become familiar with the way the bar graph is organized. Then find the facts you need.

1. Read the title of the bar graph and the labels on each axis of the graph.

2. Notice the scale, or numbers, on one of the axes.

3. Find a specific fact on the graph by reading up or down one axis and across the other axis.

Exercise 2: Use the bar graph to answer the questions.

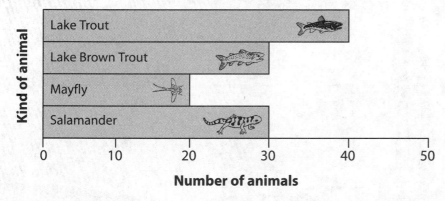

Animals Found in Lake

1. What do the two axes on the graph represent? Which has a scale?

_____

_____

2. Does the lake have more salamanders or does it have more lake trout?

_____

3. How many mayflies were found in the lake?

_____

Check your answers on page 167.

acid rain
any rain, snow, or sleet that is acidic

pollutant
something that pollutes, especially a waste material that affects air, soil, or water

react
when chemicals change each other

acid
a substance that has a pH below 7

Pollution is a fact of life in our world. **Acid rain** is a term you've probably heard in connection with pollution.

The pollution that causes acid rain comes from cars, factories, and power plants. They all burn fuels, which include coal, gasoline, and oil. When fuels are burned, **pollutants** that contain sulfur and nitrogen are given off. In the air, these pollutants **react** with gases and with rainwater to form sulfuric acid and nitric acid. This **acid** falls to the ground in acid rain, snow, sleet, and hail.

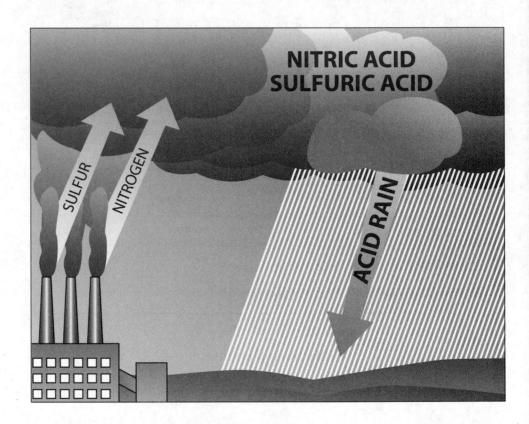

Acid rain forms when pollutants mix with water vapor in the clouds. This causes rain to be acidic.

There are many kinds of acids. Sulfuric acid is strong and harmful. Other acids are not harmful. For example, the sharp taste of an orange is caused by citric acid.

Scientists use the **pH scale** to show how strong an acid or **base** is. The pH scale runs from 0 to 14. Acids have a pH of less than 7. Bases have a pH of more than 7. Bases often feel slippery to the touch. Examples of common bases are detergents, soap, and baking soda. Liquids are either an acid or a base, except for pure water. Pure, or distilled, water is neither an acid nor a base. It has a pH of 7.

The strongest bases have a pH near 14 on the pH scale. The strongest acids have a pH near 1. Acid rain has a pH of less than 5.6. Acid rain in some parts of the United States and Europe has a pH of 2 or 3. That's more acidic than vinegar.

pH scale
a measure of how strong an acid or base is

base
a substance that has a pH between 7 and 14

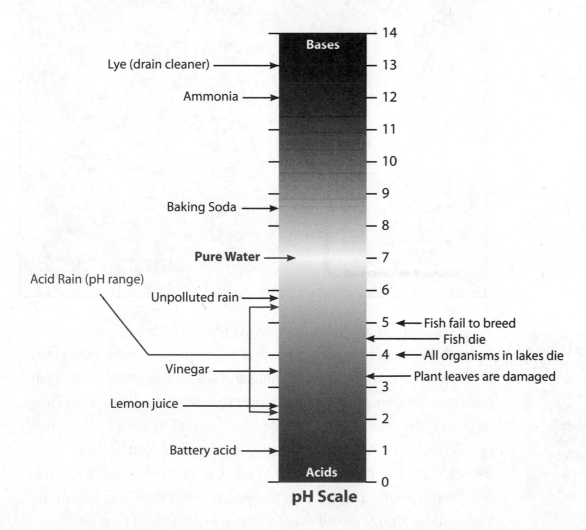

pH Scale

Acid rain can damage stone and metal. It causes chunks of stone to crumble and fall off. As a result, it damages buildings, monuments, and tombstones. It eats away at metal, causing car finishes to pit and rust.

Acid rain damages and kills plants and animals. When acid rain falls on lakes and rivers, it makes them acidic. This has already killed all life in thousands of lakes. More than 15 million acres of forest around the world are dead or dying because of acid rain falling on the trees.

Acid Rain Damage Around the World

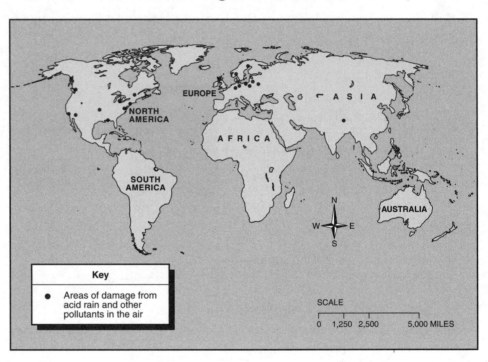

You might think that acid rain occurs only where fuels are burned. Most acid rain does fall near cities and factories. However, acids in the air can be carried thousands of miles by the wind. That means acid rain can fall even in areas that don't have any factories or power plants. Recently, people have become more aware of this problem. Efforts are being made to control the pollution that causes acid rain. If people use less electricity, power plants will use less fuel. If people drive their cars less, there will be less pollution. Doing these things would lead to less acid rain. Acid rain is a problem that is not easy to solve. Each person has to help reduce the amount of pollution in the air.

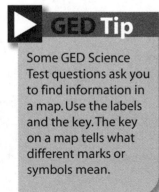

▶ GED Tip

Some GED Science Test questions ask you to find information in a map. Use the labels and the key. The key on a map tells what different marks or symbols mean.

Practice

Vocabulary ■ Write the word that best completes each sentence.

1. The damage caused by acid _____ includes rusting metal and crumbling stone.

2. A _____ has a pH greater than 7.

3. An _____ has a pH less than 7.

acid

base

pH

rain

Cause and Effect ■ Circle the number of the correct answer.

4. Acid rain forms because

    (1) vinegar is in the air.

    (2) pollutants in the air react with rainwater.

    (3) pure water has a pH of 7.

    (4) the rain has a pH of 5.6.

5. Effects of acid rain include

    (1) citric acid in lakes.

    (2) the burning of fuel.

    (3) the killing of plants and animals.

    (4) the forming of bases in lakes.

Reading a Map ■ Use the map on page 108 to choose the answer. Circle the number of the correct answer.

6. Damage has been caused by acid rain in

    (1) North America and South America.

    (2) Africa and Asia.

    (3) North America and Europe.

    (4) South America and Australia.

7. Based on this map and the paragraph after it, it is likely that Europe has

    (1) many cities with heavy traffic.

    (2) small amounts of rain.

    (3) many factories.

    (4) both (1) and (3)

Check your answers on page 167.

Combining Words and Pictures

On the GED Science Test, you answer questions based on information that you learn from both words and pictures.

One strategy for passing the GED Test is to practice combining the information given in words with the information in a picture.

 Strategy Try the strategy on the example below. Use these steps.

Step 1 Look at the picture. Read the title and labels.

Step 2 Read the paragraph. What new information does the paragraph give you that you could add to the information in the picture?

Step 3 Answer the question.

Example

Catalytic converters are found on cars. Harmful pollutants made by the car's engine go into the catalytic converter. Materials in the catalytic converter cause a chemical reaction that turns the pollutants into water, nitrogen, and carbon dioxide—materials that are naturally found in the air.

Which substance comes out of the exhaust pipe?

(1) water

(2) chemical reaction

(3) nitrogen oxide

(4) carbon monoxide

Catalytic Converter

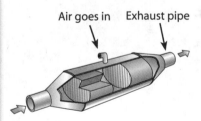

Air goes in   Exhaust pipe

Harmful
pollutants
(carbon monoxide
nitrogen oxide)

In Step 1 you learned that the picture shows a catalytic converter. In Step 2 you read that catalytic converters change harmful pollutants into less harmful substances. In Step 3 you read the question. You found the exhaust pipe in the picture. You combined the information about materials leaving the catalytic converter with your understanding of the picture. The correct answer is (1). Choices (2), (3), and (4) are not materials that come out of the exhaust pipe.

Practice the strategy. Use the steps you learned. Circle the number of the correct answer.

Scrubber

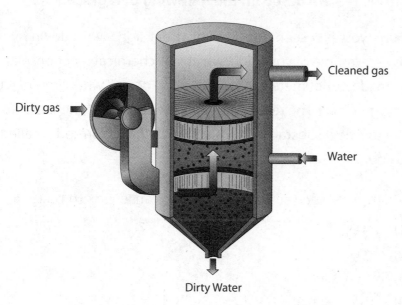

Dirty gas

Cleaned gas

Water

Dirty Water

Many factories have scrubbers. A scrubber takes harmful pollutants out of the smoke given off by a factory. Without scrubbers, the pollutants would go into the air we breathe.

1. Where are you MOST likely to find a scrubber like the one shown in the picture?

    (1) under a car
    (2) inside a house
    (3) in a weather station
    (4) inside a factory smokestack

2. Which substance does a scrubber use to clean a factory's output?

    (1) gas
    (2) air
    (3) water
    (4) smoke

Check your answers on page 167.

Unit 3 GED Test Practice

Read each paragraph and question carefully. Circle the number of the correct answer.

Questions 1–3 are based on the following paragraph.

Have you ever seen rusty cars or street signs worn down by rust? Rust is an example of corrosion, which is a chemical reaction. Many things made out of metal can rust. Rusting occurs when iron reacts with oxygen in air. The reaction causes the surface of the metal to change to a new substance. The substance that is formed is called iron oxide.

1. Which substance reacts with iron to cause cars to rust?

(1) water
(2) oxygen
(3) iron oxide
(4) carbon dioxide

2. Based on the paragraph, the word *corrosion* means

(1) wearing away.
(2) all chemical reactions.
(3) becoming stronger.
(4) restoring to the original form.

3. Which of the following is NOT likely to rust?

(1) a shovel
(2) a spoon
(3) a rubber bicycle tire
(4) a street sign

Questions 4–6 are based on the following table and paragraph.

Substance	Chemical Symbol	Melting Point* (0° in Celsius)
Mercury	Hg	-39
Oxygen	O_2	-218
Sucrose	$C_{12}H_{22}O_{11}$	185
Water	H_2O	0

*A minus sign (–) in front of a number means "below zero."

What is the melting point of a substance? It is the temperature at which the substance turns from a solid into a liquid. The temperature at which a substance changes from a liquid to a solid is called the freezing point. A substance's melting point is the same temperature as its freezing point.

4. What causes a substance to melt?

(1) heating the substance slightly
(2) removing the substance from the freezer
(3) moving the substance from a cold to a hot place
(4) bringing the substance to its melting point

5. Which substance has the highest freezing point?

(1) mercury
(2) oxygen
(3) sucrose
(4) water

6. Which substance has the lowest melting point?

(1) mercury
(2) oxygen
(3) sucrose
(4) water

Like wine, beer is the result of fermentation. Fermentation is the chemical reaction that happens when yeast get energy from sugar. The sugar found in grain is fermented by yeast, forming alcohol. The carbon dioxide gas given off during fermentation is trapped, causing beer to have bubbles in it.

7. Which of these is NOT an effect of fermentation?

   (1) the carbon dioxide in wine

   (2) the alcohol content in beer

   (3) the bubbles in beer

   (4) the sugar in grain

8. Based on the paragraph, the bubbles in beer are made of

   (1) sugar.

   (2) alcohol.

   (3) yeast.

   (4) carbon dioxide.

Check your answers on page 167.

Unit 3 Skill Check-Up Chart

Check your answers. In the first column, circle the numbers of any questions that you missed. Then look across the rows to see the skills you need to review and the pages where you can find each skill.

Question	Skill	Page
3, 8	Finding Facts	Unit 1, 16–17
2	Using Context Clues	Unit 2, 74–75
1, 4, 7	Understanding Cause and Effect	98–99
5, 6	Reading Tables and Bar Graphs	104–105

Unit 4   Physics

In this unit you will learn about

- sound
- gravity
- light
- energy

atom

gravity

decibel

spectrum

Physics is the study of how things move and change. Physics looks at our physical world to figure out how things work.

Physics explains how sound travels. Name two sounds that you have heard today.

_____

_____

Many of our modern machines and tools were invented thanks to physics. Microwave ovens and lasers are examples of things that were invented using physics.

What is a machine or tool that didn't exist one hundred years ago?

_____

_____

If a tree falls in the forest and no one is around to hear it fall, does it make a sound? The scientific answer to the question is "yes." When the tree hits the ground, it causes the tiny particles that make up the air and ground to move back and forth, or vibrate. The vibrations make **sound waves**.

sound wave
a pattern of vibrations carried through air, water, and solid objects

To see how sound waves are started by vibrations, take a rubber band and pull it tight. While it is still held tight, pluck the rubber band. You will see the rubber band vibrating, and you will hear sound. That's because the vibrations cause sound waves.

Vibrations push the air in the same way that a stone makes waves when you drop it into a pond. A small stone makes small waves that travel out to the edges of the pond. A large rock makes large waves. In the same way, a small vibration causes small sound waves.

Vibrations cause waves.

Most of the sounds we hear travel through air. However, sound can also travel through solid objects. Put your ear down on a tabletop and tap on the table. You will hear a sound. Sound travels through water, too. Have you ever been swimming underwater and heard the sound made as someone jumps in? As long as there is something for it to travel through, sound will travel.

The table lists some different materials and the speed that sound travels through them. As you can see, sound waves travel faster through solid objects than through air.

Speed of Sound	
Material	**Speed in Feet Per Second**
Air (68°F)	1,125
Water	4,915
Lucite plastic	8,793
Steel	16,600
Aluminum	16,700
Pyrex glass	18,500

If sound travels faster through solid objects, why is it easier for us to hear a person who is talking directly to us than someone who is talking through a wall? This is because sound waves, like rubber balls, bounce off walls and other surfaces and cause a **reflection**. The echoes we hear in a tunnel are the sound waves reflecting off the walls of the tunnel. But not all of the sound waves that hit a wall are reflected. Some are absorbed. Some pass through the wall. It can be difficult to hear a person talking through a wall because only a few sound waves actually pass through the wall. Most of them are reflected back to the speaker or absorbed by the wall.

Sound has two major characteristics: **volume** and **pitch**. Volume (or loudness) is measured in **decibels** (dB). Pitch (or how high or low a sound is) is measured in **hertz** (Hz).

The louder the sound, the more energy it has. The quietest sound that most people can hear has a volume of 0 decibels. A whisper has a volume of 20 decibels. The sound of a moving subway train is about 100 decibels. This sound has a much higher intensity than the quietest sound that can be heard.

Being exposed to loud sounds can damage your ears. The 100-decibel subway is loud enough to damage the ears of someone who is exposed to this sound for long periods of time. The sound of a jet engine is even louder. This is why baggage handlers and other people who work around jet engines wear ear protection. People who work around other noisy machines also wear ear protection.

reflection
the effect that happens when light, sound, or heat bounces off a surface

volume
loudness of a sound

pitch
quality of a sound

decibel
a unit of measurement of volume

hertz
a unit of measurement of pitch

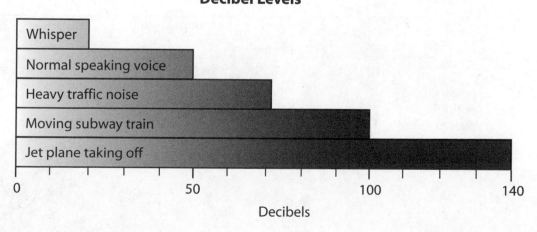

Decibel Levels

Whisper
Normal speaking voice
Heavy traffic noise
Moving subway train
Jet plane taking off

0   50   100   140
Decibels

The diagram below shows four different sounds. Each sound is made up of sound waves. Notice that the waves are different. The height of the waves is related to their **volume**. The higher the sound waves are, the louder the volume is. The distance between the waves is related to their **pitch**. The closer the waves are, the higher the pitch is.

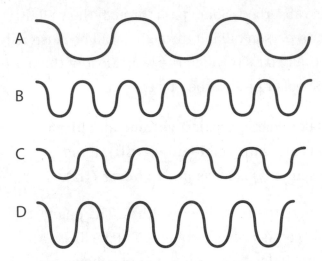

Sounds A and B have the same volume. B has a higher pitch than A, because its waves are closer together.

Sounds C and D have the same pitch. The distance between the waves is the same. D is louder than C because the height of the wave is greater.

The human ear can hear pitches from as low as 20 Hz to as high as 20,000 Hz. Bats and porpoises can make sounds that are as high as 100,000 Hz. That is five times higher than humans can hear.

The outer ear, or the part of the ear we see, receives sound waves and directs them toward the eardrum. Like a real drum, the eardrum vibrates. This transmits the sound waves to the inner ear. The inner ear is full of a liquid that is moved by the sound waves. Hairlike structures in the inner ear feel the moving liquid and send messages to the brain. When this happens, we are hearing.

▶ GED Tip

Some GED Science Test questions ask you to compare information in diagrams. The labels and captions can help you find the information you need.

How the Ear Hears

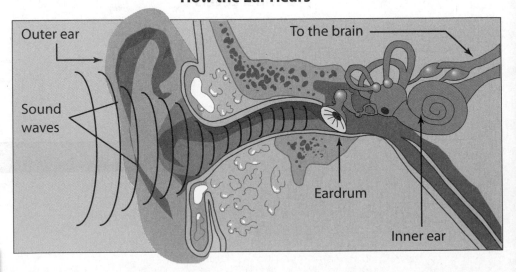

Practice

Vocabulary ■ Write the word that best completes each sentence.

1. A _____ is a way of measuring a sound's loudness.

2. _____ occurs when sound waves bounce off a surface.

decibel

hertz

reflection

Cause and Effect ■ Circle the number of the correct answer.

3. What happens when you pluck a rubber band that is pulled tight?

   (1) You see the rubber band vibrating.
   (2) You hear a sound.
   (3) You see the sound waves.
   (4) both (1) and (2)

4. What makes echoes happen?

   (1) Sound waves reflect off walls and other surfaces.
   (2) Sound travels faster through solid objects.
   (3) Some sound waves are absorbed.
   (4) both (2) and (3)

Using Context Clues ■ Circle the number of the correct answer.

5. In the third paragraph on page 117, *intensity* probably means

   (1) pitch.
   (2) energy level.
   (3) frequency.
   (4) sound wave.

6. In the last paragraph on page 118, *transmit* probably means

   (1) stop.
   (2) make noise.
   (3) absorb.
   (4) send.

Check your answers on page 168.

Reading Line Graphs

A line graph shows how two things are related. Like bar graphs, line graphs have two lines called axes. Each axis (bottom line and side line) measures something. The graph's title and axis labels tell what the graph shows.

 Strategy Look at the line graph. Ask yourself: What kind of information does the line graph show?

1. Read the graph's title and the label of each axis. What kind of information does the graph show?

2. Look at the scale, or numbers, on each axis. Ask yourself: Which two things are being related? How are they measured?

3. Look at the line. Ask yourself: How are the two things related?

Exercise 1: The line graph shows the speed of sound in air. Use the graph to answer the questions.

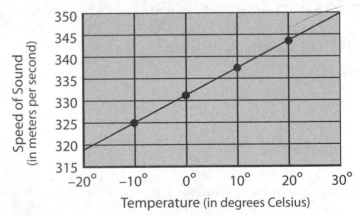

Speed of Sound in Air at Different Temperatures

1. Which two things are being measured in the graph? _____

   _____

2. Is the speed of sound greater at 0 degrees Celsius or at 20 degrees Celsius? _____

   _____

Once you understand the information in a line graph, you can use the line graph to help you make a prediction. Imagine that the line in a graph continues. What would the line tell you?

 Strategy Look at the line in the graph. What does the line tell you?

1. Read the title and the axis labels. Ask yourself: What does each axis show? What is being related?

2. Look at the shape of the line in the graph. Ask yourself: If the line continued, in what direction would it go?

Exercise 2: Look at the following line graph. Imagine that the runner keeps the same speed. How far will the runner have traveled at 5 minutes?

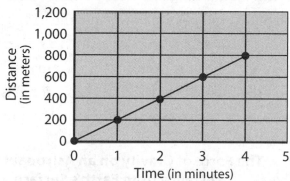

Runner Moving at Constant Speed

Exercise 3: Look at the following line graph. Kelvin is a temperature scale like Fahrenheit or Celsius. Imagine that the line continues. Will the gas's volume be greater or less at 500 kelvins?

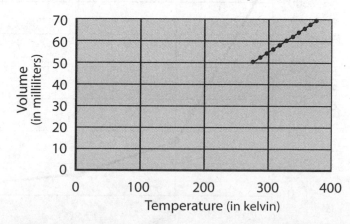

Volume of a Gas at Different Temperatures

Check your answers on page 168.

Lesson 20   Gravity

gravity
the force that pulls any two objects together

newton
a unit that measures force

If you drop a ball, it falls down. If you trip, you fall down, too. The earth's **gravity** is the force that makes these things happen. Gravity holds you on the earth's surface. This force also holds the earth, moon, and planets in their places.

Earth is not the only object that has gravity. This force exists between any two objects. The strength of the force depends on two things. The first thing is the mass of the objects. The more mass the objects have, the greater the force between them. The second thing is the distance between the objects. The closer together the objects are, the greater the force between them.

Your weight is a measure of the force of gravity. When you step on a scale, it measures the force of the pull between you and Earth. The line graph below shows how the strength of the pull changes as a person moves farther away from Earth's surface. The farther away any object is from Earth, the smaller the force of gravity is on that object. On the line graph, force is shown in **newtons** (N). One newton is slightly less than one fourth of a pound.

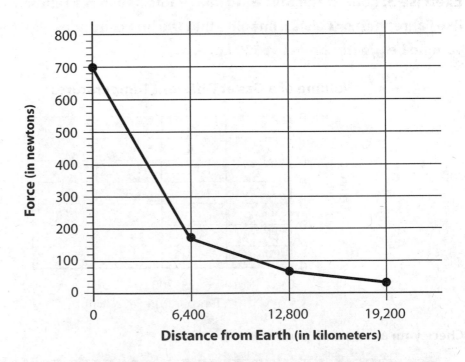

The Force of Gravity on an Astronaut and Distance from Earth's Surface

This line graph shows how the force of gravity on an astronaut changes at different distances from Earth's surface.

If you could travel to the moon, your weight would change. This is because the moon has much less mass than Earth, and therefore, the moon has only one sixth the force of gravity that is felt on Earth. An astronaut who weighs 180 pounds on Earth would weigh only 30 pounds on the moon! However, the astronaut would not be thinner or smaller. The astronaut's mass would be the same as it is on Earth. Only the pull of gravity on that mass would be different.

Earth's gravity pulls on the moon. Yet the moon does not fall to Earth's surface the way a dropped ball does. The moon is kept circling Earth by a combination of two things. The force of gravity is balanced by the moon's tendency to fly off into space.

Imagine tying a rock to a piece of string and whirling the rock around in a circle. If you let go of the string, the rock will fly off in a straight line. You can feel the rock pull on the string. The string is also pulling on the rock with something called **centripetal force**. This force pulls the rock toward the center of its circular path. When you let go of the string, nothing pulls on the rock. It is free to fly away. Gravity is the centripetal force that pulls the moon toward Earth. Earth's gravity acts like a string. It keeps the moon from flying away.

GED Tip

Pay attention to dotted lines and arrows on a diagram. Use them along with the words to figure out what the diagram is telling you.

centripetal force
the force that causes things to go toward the center

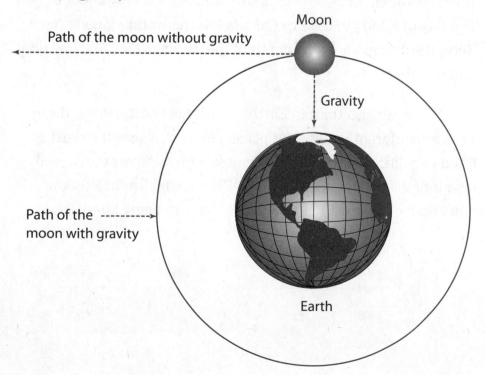

Moon

Path of the moon without gravity

Gravity

Path of the
moon with gravity

Earth

The moon stays in its orbit because the moon's tendency to move out into space is balanced by the pull of Earth's gravity.

You have seen that Earth pulls on the moon. The moon pulls on Earth, too. The moon's gravity is not strong enough to pull Earth out of its orbit. However, it does cause Earth to move slightly. The pull of the moon's gravity on Earth causes the **tides**.

The diagram below shows that high tide happens on the side of Earth facing the moon. On this side of Earth, the moon's gravity pulls the water toward the moon. A high tide also happens on the side of Earth facing away from the moon. This second high tide is also caused by the moon's gravity. The moon pulls on Earth. This pulls Earth away from the water that is on the side of Earth facing away from the moon. This creates a bulge of water on both sides of Earth. This is why we have two tides each day.

tide
the rise and fall of the level of the oceans

The moon pulls on Earth and Earth's oceans, causing the water to rise. This is called high tide.

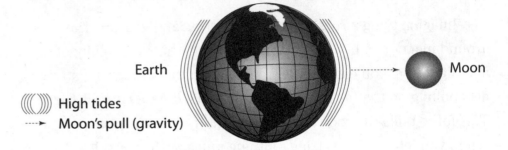

Earth

Moon

))) **High tides**
---▶ **Moon's pull (gravity)**

The parts of Earth that are not having high tides are having low tides. The change in the depth of the water between high and low tides varies. In most places, the difference is only a foot or two. The Bay of Fundy in eastern Canada has the world's largest tides. Here, the difference in the depth of the water at high and low tide is about 45 feet.

There are also tides in Earth's crust. Like ocean tides, these tides are different in different places. In most places the crust is lifted very little. However, some cities, such as Moscow, rise and fall more than 20 inches every day. People who live in Moscow don't notice the crust rising because they are being lifted, too.

Practice

Vocabulary ■ **Write the word or words that best complete each sentence.**

1. Earth's _____ causes the _____ that keeps the moon from spinning out into space.

2. Forces are often measured in _____.

| centripetal force |
| gravity |
| newtons |
| rotates |

Reading Line Graphs ■ **Use the line graph to answer each question. Circle the number of the correct answer.**

3. What would the force of gravity be on an astronaut who is 12,800 kilometers above Earth's surface?

   (1) 12,800 N
   (2) 720 N
   (3) 80 N
   (4) 180 N

4. How does the force of gravity on the astronaut compare at Earth's surface and at a distance of 19,200 kilometers?

   (1) The force is the same.
   (2) The force is less at Earth's surface.
   (3) The force is about 675 N more at 19,200 kilometers.
   (4) The force is about 675 N more at Earth's surface.

The Force of Gravity on an Astronaut and Distance from Earth's Surface

Cause and Effect ■ **Circle the number of the correct answer.**

5. What would happen if Earth had no gravity?

   (1) The moon would fall into Earth.
   (2) The moon would move out into space.
   (3) The moon would stay where it is.
   (4) The moon would move faster.

Check your answers on page 169.

Light and sound have some things in common. Both sound and light travel in waves. But light waves can travel through a vacuum or outer space—places with no air. Sound waves cannot do this. If light waves could not travel through space, we wouldn't be able to receive any light from the sun.

Light has two basic characteristics: brightness and color. A bright light, like a loud sound, has a high wave. A dim light has a smaller wave.

Light of different colors has different **wavelengths**. The wavelength tells if a wave is long or short. This is similar to the way pitch is measured in sound waves. Blue and violet light have the shortest wavelengths of all the colors. Red light has the longest wavelength. The wavelengths of yellow and orange light are in the middle. White light such as sunlight contains all the different wavelengths. In other words, white light contains all the colors of light. These colors form the **spectrum**.

When white light passes through a **prism**, the prism spreads out the colors. We can see each color separately. The prism bends the light waves. Some of the light waves bend more than others. Violet light waves bend the most. Red light waves bend the least.

wavelength
the distance between the top of one wave and the top of the next

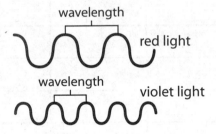

Wavelengths of different colored lights

spectrum
the different colors contained in white light: red, orange, yellow, green, blue, indigo, and violet

prism
a piece of clear glass that bends white light, showing the spectrum

When white light passes through a prism, the prism spreads out the colors, and we can see each one separately.

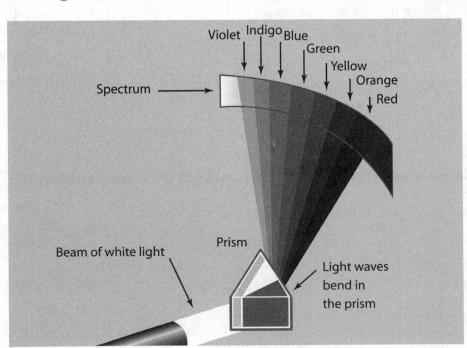

Scientists can use a special kind of prism to learn about stars. These special prisms are called **spectrographs**. A scientist who studies the stars can pass the light of a star through a spectrograph. The spectrograph separates the colors in the light, giving information about what elements are in the star.

Each element in a star absorbs light of particular wavelengths and colors. Each element has a specific "signature"—a specific set of colors. A scientist who studies stars looks at what colors and how much of each color the light from a star has. Then the scientist can figure out which elements the star contains.

spectrograph
a tool used to separate the colors of light given off by stars

laser
a device that makes a thin beam of light made up of all the same wavelength

optical fiber
thin fiber of glass or plastic used to send information

Lasers and Their Uses

Laser light is made of light waves of all the same color, or wavelength. The light waves are also organized so that they are in step with each other. **Lasers** have many everyday uses. Lasers are used

- to scan bar codes at stores and libraries.
- in surgery to correct eyesight.
- to store and read information on CDs and DVDs.
- in laser printers to print materials from a computer.
- to send telephone messages and other information through **optical fibers**.

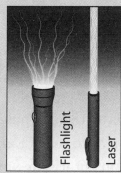

Flashlight: many different wavelengths

Laser: all the same wavelength

Flashlight

Laser

Sometimes raindrops act like tiny prisms. The raindrops break up sunlight into spectrums called rainbows. Rainbows are formed when the sun is behind you and the rain is in front of you.

Have you ever noticed during a thunderstorm that you see the lightning a few seconds before you hear the thunder? Lightning and thunder happen at the same time. However, the light reaches you much faster than the sound waves do. The speed of light is 186,000 miles per second. The speed of sound in air is 1,128 feet per second. That's only about one fifth of a mile per second.

You can use this difference to guess how far away a storm is. When you see lightning, count the seconds until you hear the thunder. It takes five seconds for sound to travel one mile, so if you count five seconds, you know that the lightning was one mile away.

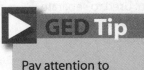

GED Tip

Pay attention to numbers. The numbers in a diagram help you understand exactly what the diagram is telling you.

How Far Away Is The Storm?

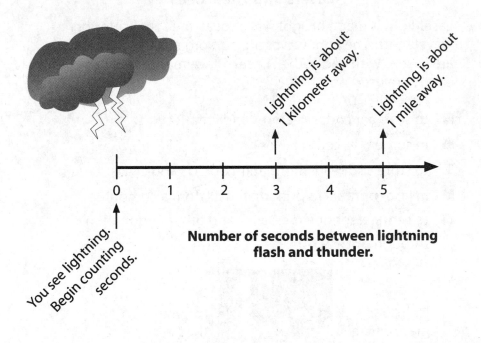

Lightning is about 1 kilometer away.

Lightning is about 1 mile away.

You see lightning. Begin counting seconds.

Number of seconds between lightning flash and thunder.

Even though light travels at 186,000 miles a second, it takes many years for the light of a star to reach Earth. For this reason, scientists measure distances in outer space in **light years**. A light year is almost 6 **trillion** miles. The closest star to the earth is four and one-half light years away. Some stars are 80,000 light years away. That means that the light from the stars we see at night left those stars 80,000 years ago.

light year
the distance that light travels in one year

trillion
1,000,000,000,000

Practice

Vocabulary ■ **Write the word that best completes each sentence.**

1. A _____ separates white light into different colors.

2. A light year is a distance of about 6 _____ miles.

3. The different colors that make up white light are called the _____ .

4. Red light has a longer _____ than violet light.

laser
prism
spectrum
trillion
wavelength

Reading Diagrams ■ **Use the diagram on page 126. Circle the number of the correct answer.**

5. In order to see a rainbow, we need

   (1) a glass prism.
   (2) white light.
   (3) raindrops.
   (4) both (2) and (3)

Cause and Effect ■ **Circle the number of the correct answer.**

6. Lightning and thunder happen at the same time. Why do you usually see lightning before you hear thunder?

   (1) Sound waves travel faster than light waves.
   (2) Light waves travel faster than sound waves.
   (3) Light contains many different wavelengths.
   (4) Sound contains many different wavelengths.

7. When white light passes through a prism,

   (1) some of the light waves bend more than others.
   (2) the light waves disappear.
   (3) we can see each color of the spectrum separately.
   (4) both (1) and (3)

Check your answers on page 169.

GED Skill Strategy

Drawing Conclusions

Sometimes you have to think beyond what is directly written in a paragraph. If you study the facts and think about what is being said, you can get a lot of information from a paragraph. You can use all of this information to draw your own conclusions. A conclusion is a judgment you can make after studying all of the facts. Read the following example.

> A small vibration causes small sound waves.
> A large vibration causes large sound waves.

From these sentences, you can conclude that the size of a sound wave depends on the strength of the vibration that caused it.

 **Strategy** Study the facts. Then think about them.

1. As you read a passage, ask yourself: What are the facts?
2. Then look for ideas you can figure out.
3. Ask yourself: What do the facts and ideas in the passage tell me?

Exercise 1: Read the paragraph. Underline the facts.

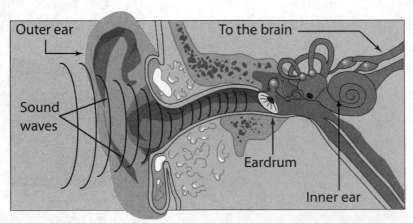

Parts of the ear

How do we hear? The outer ear receives sound waves and directs them toward the eardrum. The eardrum vibrates and passes the sound waves to the inner ear. The inner ear contains a liquid that is moved by the sound waves. Hairlike structures in the inner ear sense the moving liquid and send messages to the brain. These messages are what we call hearing.

Exercise 2: Read the paragraph on page 130 again. What do all the facts about the ear have in common? Circle the number of the correct answer.

    (1)   They are all about how the brain works.

    (2)   They are all about how hearing occurs.

    (3)   They are all about liquids.

    (4)   both (1) and (3)

Exercise 3: From the paragraph and diagram on page 130, what conclusion can you draw about the purpose of the ear? Circle the number of the correct answer.

The purpose of the ear is to

    (1)   make sounds so you can hear.

    (2)   collect sounds so you can hear.

    (3)   block loud sounds so you can't hear them.

    (4)   both (1) and (3)

Exercise 4: Read the paragraph. Then answer the questions.

      We all live with gravity. It's part of our lives. Because of gravity, when we trip, we fall down and not up. Because of gravity, people, buildings, trees, and the oceans do not fly off into space while the earth turns.

1.   What do you think would happen if there were no gravity on the earth?

    (1)   Dropped objects would fall slowly.

    (2)   Objects would fly off into space.

    (3)   Thrown objects would fall down more quickly.

    (4)   both (1) and (2)

2.   What facts from the paragraph helped you draw this conclusion?

    A. _____

    B. _____

Check your answers on page 169.

Lesson 22   Energy

energy
the ability to change or move things

atom
the smallest part of matter

nucleus
an atom's center that contains protons and neutrons

Energy is the ability to change or move things. It is in the sunlight that warms us and lights our planet. It is in the food that we eat and in the trees that we cut down and burn. There is energy in every **atom** of matter, even if the atom does not appear to be moving.

Energy comes in many forms. Here are a few of the most important kinds of energy.

- Solar energy comes from the sun, in the forms of heat and light.

- Electrical energy comes in many forms, such as batteries, lightning, and the wires in a power line.

- Chemical energy is stored in many ways, such as in our food and in the cells of our bodies. When chemical bonds are broken and new bonds are formed, energy may be released.

- Nuclear energy gets its name from the **nucleus**, the center of the atom. Nuclear energy is produced in two ways. One way is to join the nucleus of one atom with the nucleus of another atom. This is the way energy is produced in the sun. A second way is to split an atom and turn a tiny bit of matter into a huge amount of energy. This kind of energy is produced in nuclear power plants.

Forms of Energy

Sun
(Solar Energy)

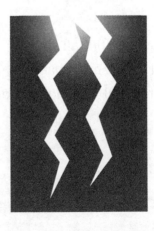

Lightning Bolt
(Electrical Energy)

Apple
(Chemical Energy)

Energy cannot be created or destroyed. It can only change its form or pass from one object to another object. Solar energy can **transform** into chemical energy. Nuclear energy can transform into electrical energy. No energy is lost.

However, in many energy transfers some energy turns into waste heat. When you hit a ball with your hand, not all of the energy you produce is passed to the ball. Your hand stings and feels hot after you hit a ball because some heat energy has passed to your hand.

Every kind of energy can transform into other kinds of energy. A **nuclear power plant** transforms nuclear energy into electrical energy. Power lines carry the electrical energy to your toaster. Your toaster transforms that electrical energy into the heat energy needed to toast your bread. Your body can then transform the chemical energy in the toast into the energy you need to breathe, move, and live.

transform
change into another form

nuclear power plant
factory that changes energy from the nucleus of an atom into electrical energy

Energy Transformation

Nuclear energy from a power plant is eventually transformed into energy our bodies use.

A single machine like a car engine may use several energy transformations. In a car engine, electrical energy makes a spark. The heat of the spark releases chemical energy in the fuel. The chemical energy produces greater heat. The heat makes some of the small car parts move up and down, making larger car parts turn, and finally making the car tires spin.

The sun provides most of the energy on Earth. The energy arrives as heat and light. As soon as the heat and light reach Earth, energy transformations begin.

Plants transform light into chemical energy for themselves. Any animal that eats the plants is also able to use that chemical energy. Plants can also be burned to release heat and light. A fuel made from plant material or animal waste is called a **biomass fuel**.

biomass fuel
fuel made from plant material or animal waste

fossil fuel
coal, oil, gas, and other fuels that formed from the bodies of living things millions of years ago

Fossil fuels also store energy from the sun. Millions of years ago, plants and animals used energy from the sun. After they died, their remains mixed with sand, rock, and mud. Over millions of years these remains slowly changed into coal, oil, and natural gas. These are the three major fossil fuels. The energy from these fuels started out as energy from the sun.

Scientist Albert Einstein showed how energy and matter are related in the famous equation $E = mc^2$. E is energy, m is matter, and c^2 is the speed of light multiplied by itself. When matter is transformed into energy, the amount of energy produced is enormous. Even matter is a form of energy!

Power plants use fossil fuels or nuclear energy for power.

Practice

Vocabulary ■ Write the word or words that best complete each sentence.

1. The center of an atom is called the _____ .

2. Coal, oil, and natural gas are all _____ .

biomass fuels

fossil fuels

nucleus

Drawing Conclusions ■ Write your answer below.

3. According to the list on page 132, what kind of energy is used in batteries?

_____

4. According to the list on page 132, what kind of energy is stored in an apple?

_____

Reading a Line Graph ■ Use the graph on the right to answer each question. Circle the number of the correct answer.

5. When a material gains heat energy, it melts, then boils. Which point shows the most heat energy?

   (1) A
   (2) B
   (3) C
   (4) D

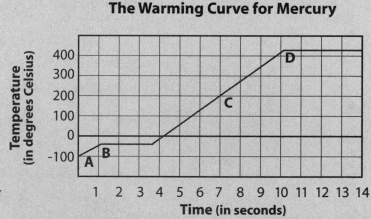

The Warming Curve for Mercury

6. How many seconds does it take for mercury to go from 0 degrees Celsius to 300 degrees Celsius?

   (1) 5 seconds
   (2) 10 seconds
   (3) 300 seconds
   (4) 357 seconds

Check your answers on page 170.

GED Skill Strategy

Predicting Outcomes

An outcome is a type of effect. It is something that will happen as a result of something else that happens. When you predict, you make an informed guess about what is likely to happen. You base your prediction on what you already know.

 Strategy   Review what you already know. Then apply it to the new situation.

Step 1   Think about what has happened before in similar situations.

Step 2   Think of facts that you can apply to this situation.

Step 3   Ask yourself: What is the best guess I can make based on what I know?

Exercise 1: Read the text. Then study the diagram to find out how rainbows form.

Raindrops can act like tiny prisms. The raindrops break up sunlight into rainbows. The diagram below shows that rainbows are formed when the sun is behind you and the rain is in front of you.

You see a rainbow when raindrops bend sunlight and separate it into different colors.

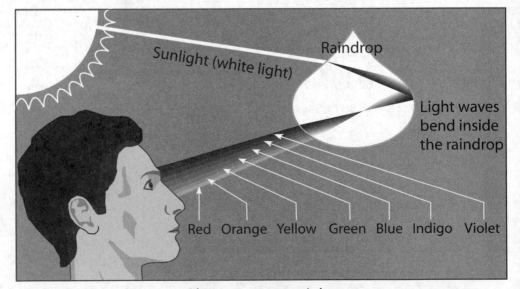

Observer sees a rainbow

The text and diagram give the following facts.

- Sunlight can be broken up by raindrops into rainbows.

- You see a rainbow when the sun is behind you and the rain is in the sky in front of you.

Predict whether you are likely to see a rainbow in each of the following situations. Explain your predictions. Write your answers.

1. You are watching the sun rise above a faraway hill. As you watch you realize that it has started raining where you are.

   _____

   _____

2. The sky has been completely covered with clouds all day. You set out for a walk anyway and get caught in the rain.

   _____

   _____

Exercise 2: Read the paragraph. Then answer the question.

Why is the sky blue? White light from the sun passes through our atmosphere. It then becomes scattered by tiny particles such as dust and pollen. These particles are big enough to reflect blue light, but not big enough to reflect much red light. This is because red light has a longer wavelength. When the sun is rising or setting, the light must pass through many more particles, so we see more red and orange at these times.

3. Your friend predicts that if a volcano erupted and put a lot of extra dust into the air, the sky would look red in the daytime. Do you agree with this prediction? Explain your answer.

   _____

   _____

Check your answers on page 170.

Choosing the Right Answer

On the GED Science Test, you have to choose the right answer from several choices.

One strategy for choosing the right answer is to read all of the answer choices before you decide whether any one answer is correct or not. Then think about each answer choice, and decide which one is correct.

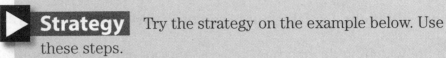 **Strategy** Try the strategy on the example below. Use these steps.

Step 1 Read the question and all the answer choices. It is important to read all four choices even if you think you have found the correct answer right away.

Step 2 Cross out the choices with facts you know to be incorrect.

Step 3 Cross out choices that do not answer the question.

Step 4 Choose an answer. Check to make sure that the answer makes sense.

Example

Sometimes only one energy transformation is needed. Sometimes more than one is needed. For example, an apple changes the sun's light directly into the chemical energy the apple needs. A person has to eat the apple to get the apple's chemical energy. Then the person's body changes the chemical energy into the different kinds of energy a human needs.

What is the main idea of the paragraph?

(1) Energy from the sun has many uses.

(2) An apple and a person use energy differently.

(3) Some energy changes have more steps than others.

(4) Energy from the sun always ends as nuclear energy.

In Step 1 you read the question and the four choices. In Step 2 you crossed out choices with incorrect facts, like choice (4). In Step 3 you crossed out choices that do not answer the question, such as choices (1) and (2). In Step 4 you reread the question and choice (3) to make sure the answer made sense. Choice (3) is the correct answer.

Practice

Practice the strategy. Use the steps you have learned. Circle the number of the correct answer.

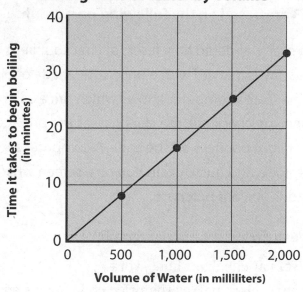

Boiling Time of Water by Volume

Time it takes to begin boiling (in minutes)

Volume of Water (in milliliters)

1. What is the main idea of the graph?

   (1) Greater volumes of water take more heat to boil.

   (2) Greater volumes of water take more time to boil.

   (3) It takes 10 minutes for 500 milliliters of water to boil.

   (4) It takes about 20 minutes to boil a large pan of water.

Practice the strategy. Use the steps you have learned. Circle the number of the correct answer.

Laser light differs from other kinds of light in the ways the waves line up. In other lights, the tops and bottoms of the waves are not lined up. In a laser light, the tops and bottoms of the waves do line up. A laser light is said to be *in phase*.

2. Based on the paragraph, which of the following is true?

   (1) Laser light is in phase.

   (2) Laser light is out of phase.

   (3) The waves in laser light are all mixed up.

   (4) The waves in most kinds of light are lined up.

Check your answers on page 170.

GED Test Practice

Read each paragraph and question carefully. Circle the number of the correct answer.

Questions 1–3 are based on the following paragraph.

Have you ever wondered how lasers at checkout lines work? Each product in the store is marked with a price code made of black and white bands. The store's computer knows which price each code stands for. When you check out, the clerk passes the price code across a laser. The laser light bounces off the code. A computer senses the light, reads the code, and quickly tells you on a screen what each item costs. Then it adds up your purchases.

1. What is the main idea of the paragraph?

   (1) how lasers at checkout lines work
   (2) how a store clerk passes the price code across a laser
   (3) how a store's computer adds up customer purchases
   (4) how laser lights bounce off price codes

2. Predict what would happen if the price code on a product were rubbed out.

   (1) The store's computer would add the price to your purchases.
   (2) The store's computer would not read the price code.
   (3) The store's clerk would have to use a different laser to read the price code.
   (4) The store's clerk would have to use a different computer to add up the purchases.

3. Lasers are used at checkout lines because

   (1) the clerks prefer the lasers.
   (2) lasers make the checkout process faster.
   (3) the computer never makes a mistake.
   (4) lasers make it harder for people to shoplift.

Questions 4–6 are based on the following graph.

A Comparison of Weight and Mass

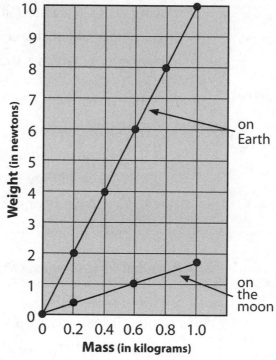

4. Which conclusion can you draw based on the information in the line graph?

   (1) Mass and weight are the same thing.
   (2) Mass is always a larger number than weight.
   (3) An object weighs the same on Earth as on the moon.
   (4) An object has the same mass on Earth as on the moon.

5. An object that weighs 1 newton on the moon will weigh

   (1) 0.6 newton on Earth.
   (2) 1 newton on Earth.
   (3) 6.5 newtons on Earth.
   (4) 10 newtons on Earth.

6. An object has a mass of 1.2 kilograms. Predict its weight in newtons on Earth.

   (1) about 1.2 newtons
   (2) about 2 newtons
   (3) about 8 newtons
   (4) about 12 newtons

Questions 7–8 are based on the following paragraph.

Your power company can get energy to make electricity from moving water in a waterfall. Your power company could also get energy by burning a fuel, such as coal or oil. Some power companies use nuclear energy, which is the energy given off when atoms are split.

7. What is the main idea of the paragraph?

   (1) Power companies get energy from different sources.
   (2) Most power companies get energy from waterfalls.
   (3) Nuclear energy is the most dangerous source of electricity.
   (4) Your power company probably makes electricity by burning coal.

8. What kind of energy results when atoms are split?

   (1) fuel energy
   (2) nuclear energy
   (3) water energy
   (4) electrical energy

Check your answers on page 170.

Unit 4 Skill Check-Up Chart

Check your answers. In the first column, circle the numbers of any questions that you missed. Then look across the rows to see the skills you need to review and the pages where you can find each skill.

Question	Skill	Page
1,7	Understanding the Main Idea	Unit 1, 40–41
8	Understanding Cause and Effect	Unit 3, 98–99
5	Reading Line Graphs	120–121
3,4	Drawing Conclusions	130–131
2,6	Predicting Outcomes	136–137

Science Posttest

This posttest will give you an idea of how well you've learned to understand science content using the skills in this book.

You will read short science passages and graphics such as graphs, maps, and diagrams. You will also answer multiple-choice questions. There is no time limit for this test.

Read each selection and question carefully. Circle the number of the correct answer.

Questions 1–3 are based on the following paragraph and diagram.

Every year trees produce a double layer of wood. They produce a light-colored layer in the spring. Then later, in the summer, they produce a darker-colored layer. If you cut down a tree and look at the stump, you will see the rings that the layers form. Each pair of light and dark rings shows one year that the tree has lived. The older rings are on the inside, and the younger rings are on the outside. Rings vary in thickness. For example, rings that grow in rainy years are thicker than rings that grow in dry years.

1. Where are the oldest tree rings found?

    (1) in the tree's bark
    (2) in the tree's branches
    (3) in the center of the tree trunk
    (4) in the outside part of the tree trunk

2. About how old is the tree shown in the diagram?

    (1) 1 year
    (2) 5 years
    (3) 10 years
    (4) 20 years

3. What is the main idea of this paragraph?

    (1) The oldest rings are on the inside of a tree.
    (2) The rings on a tree vary in thickness.
    (3) Every year trees produce a double layer of wood.
    (4) Trees produce dark-colored rings in the summer.

Questions 4–6 are based on the following paragraph and diagram.

The amount of daylight affects the way plants grow. Short-day plants begin to flower in autumn, when the days are short. Long-day plants begin to flower in summer, when the days are long. When you try to grow long-day plants during a short-day season, they do not grow well. Similarly, when you grow short-day plants during a long-day season, they do not grow well.

Long-day plant   **Short-day plant**

  Geranium     Chrysanthemum

Grown during   Grown during   Grown during   Grown during
long days   short days   short days   long days

4. When do long-day plants start flowering?

   (1) fall
   (2) winter
   (3) autumn
   (4) summer

5. When does a geranium bloom?

   (1) during the short days of winter
   (2) during the long days of summer
   (3) during the short days of autumn
   (4) at any time of year

6. What is the main idea of the paragraph?

   (1) Short-day plants flower when days are short.
   (2) Long-day plants flower when days are long.
   (3) Short-day plants do not grow well during long days.
   (4) The amount of daylight affects how plants grow.

Questions 7–9 are based on the following map.

High Temperatures and Weather for January 22

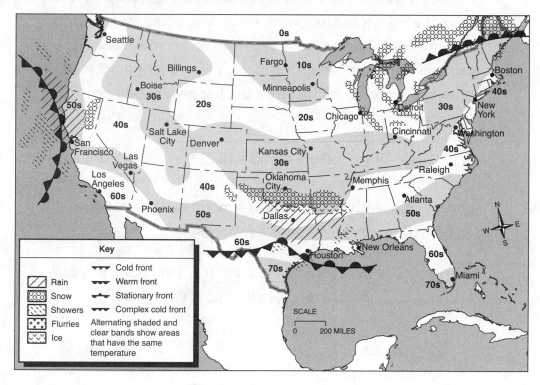

7. In which of the following cities is it raining?

    (1) Minneapolis
    (2) Denver
    (3) Atlanta
    (4) Dallas

8. What is the high temperature in Chicago?

    (1) in the 10s
    (2) in the 20s
    (3) in the 30s
    (4) in the 40s

9. What is the weather like in San Francisco?

    (1) rainy, in the 50s
    (2) rainy, in the 40s
    (3) snowing, in the 30s
    (4) clear, in the 20s

Questions 10–12 are based on the following passage.

An eclipse happens when three objects in space line up, causing one of the objects to make a shadow on one of the other objects. When the earth lines up in between the sun and the moon, the earth blocks sunlight from hitting the moon. This causes a shadow on the moon, called a lunar eclipse. When this happens, the moon is not visible from the earth.

Another kind of eclipse blocks sunlight from hitting the earth. When the moon comes in between the earth and the sun, it keeps sunlight from reaching the earth. The moon blocks the view of the sun from the earth. This is called a solar eclipse.

10. Based on the information in the first paragraph, you can conclude that the word *eclipse* probably means

    (1) circle.
    (2) shadow.
    (3) bright spot.
    (4) banana shape.

11. What cannot be seen from the earth during a lunar eclipse?

    (1) the moon
    (2) the sun
    (3) all sunlight
    (4) the other planets

12. The word *solar* in the second paragraph probably refers to

    (1) the sun.
    (2) the moon.
    (3) the earth.
    (4) all objects in space.

Questions 13–14 are based on the following paragraph and bar graph.

There are more than 110 different kinds of matter. Each kind of matter is called an element. Each element has different characteristics, or properties. A single element, such as gold, silver, or aluminum, has a set of properties that are different from the properties of any other element. When elements combine, the new combined material has different properties. Water is made of hydrogen and oxygen. Your body is made mostly of the elements oxygen, carbon, and hydrogen.

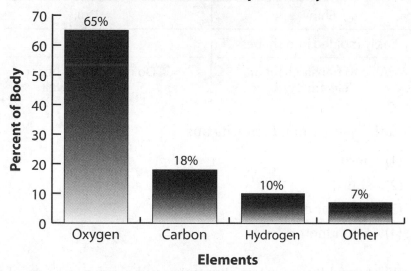

Elements that Make Up the Body

13. What causes silver and gold to have different colors?

    (1) their properties
    (2) their shapes
    (3) their temperatures
    (4) the amount of water in them

14. According to the bar graph, which element is found in the largest percentage in the human body?

    (1) carbon
    (2) oxygen
    (3) hydrogen
    (4) aluminum

Questions 15–16 are based on the following paragraph and chart.

Metals have many of the same properties, or characteristics. First, they all have luster, or are shiny. Second, all metals except mercury are solid at room temperature. Third, metals are conductors of electricity. They allow electricity to pass through them. Metals are also good conductors of heat, so most pots and pans used for cooking are made of aluminum or copper.

Properties of Metals and Nonmetals	
Metals	**Nonmetals**
Shiny	Dull
Easily molded into shapes	Brittle
Able to transfer heat or electricity	Do not transfer heat or electricity well

15. In this paragraph, *luster* means

    (1) hard.
    (2) shiny.
    (3) used for cooking.
    (4) a conductor.

16. What would you expect aluminum and copper to be used for?

    (1) insulation
    (2) safety shoes
    (3) electrical wires
    (4) potholders

Questions 17–19 are based on the following paragraph and line graph.

When two surfaces rub together, there is friction. Friction slows things down and makes them work harder. When the parts of a machine rub together, they cause friction. Friction makes machines wear out and wastes energy. You can reduce friction in a machine by adding lubrication, or a liquid that makes things slippery, to the parts.

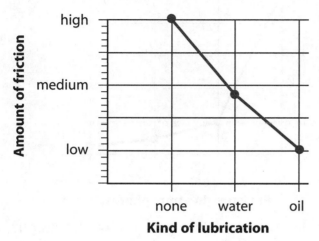

Friction Between Two Metals

17. Which type of lubrication most reduces friction?

(1) oil

(2) water

(3) none

(4) oil and water together

18. If you add water to two pieces of metal to reduce friction, the amount of friction will be

(1) none.

(2) low.

(3) medium.

(4) high.

19. You can conclude that keeping machines oiled properly

(1) wastes energy.

(2) makes them last longer.

(3) causes more friction.

(4) makes them cost more to run than machines that are not oiled.

Questions 20–22 refer to the following paragraph and graph.

If the temperature remains the same, the volume of a gas decreases when the pressure on it increases. If the temperature remains the same, the volume of a gas increases when the pressure on it decreases. This relationship is shown in the graph.

Relationship of Volume and Pressure of a Gas When Temperature Remains the Same

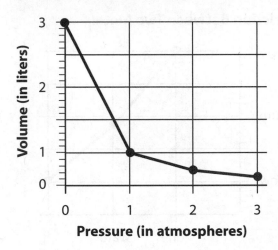

20. What is the unit of measure for pressure on this graph?
    (1) volume
    (2) gas
    (3) liter
    (4) atmosphere

21. When a gas has a volume of 1 liter, what is the pressure according to the graph?
    (1) 0 atmosphere
    (2) 1 atmosphere
    (3) 2 atmospheres
    (4) 3 atmospheres

22. What effect does increasing pressure have on a gas?
    (1) The gas takes up more space.
    (2) The gas takes up less space.
    (3) The gas takes up the same amount of space.
    (4) The gas has a volume of 1 liter.

When you finish the *Science Posttest*, check your answers on page 171. Then look at the chart on page 151.

Skills Review Chart

This chart shows you which skills you should review. Check your answers. In the first column, circle the number of any questions you missed. Then look across the row to find out which skills you should review as well as the page numbers on which you find instruction on those skills. Compare the items you circled in the *Skills Review Chart* to those you circled in the *Skills Preview Chart* to see the progress you've made.

Questions	Skill	Pages
1, 4, 11	Finding Facts	16–17
2, 5	Reading a Diagram	26–27
3, 6, 19	Understanding the Main Idea	40–41
7, 8, 9	Reading a Map	64–65
10, 12, 15	Using Context Clues	74–75
13, 22	Understanding Cause and Effect	98–99
14	Reading Tables and Bar Graphs	104–105
17, 20, 21	Reading Line Graphs	120–121
16, 19	Drawing Conclusions	130–131
18	Predicting Outcomes	136–137

Glossary

260°F, –280°F a temperature of 260 degrees Fahrenheit; a temperature of minus 280 degrees Fahrenheit. *page 60*

acid a substance that has a pH below 7. *page 106*

acid rain any rain, snow, or sleet that is acidic. *page 106*

algae a very simple type of plantlike life that usually grows in water. *page 50*

antibiotic a drug that stops the growth and reproduction of bacteria. *page 37*

antibodies chemicals made by the immune system to destroy a particular type of germ. *page 37*

arteries blood vessels that carry blood away from the heart. *page 22*

atmosphere all the air surrounding Earth. *page 60*

atom the smallest part of an element. *pages 91 and 132*

atomic number the number of protons in an atom. *page 91*

atomic structure the number of protons, neutrons, and electrons that make up an atom. *page 92*

axis an imaginary line from the North Pole to the South Pole. *page 78*

bacteria (plural of bacterium) tiny one-celled organisms. *pages 18 and 36*

base a substance that has a pH between 7 and 14. *page 107*

Big Bang the explosion that may have started the universe. *page 76*

Big Dipper a constellation made up of seven stars. *page 81*

biomass fuel fuel made from plant material or animal waste *page 134*

blood vessels tubes that blood flows through. *page 22*

bovine growth hormone (bGH) a substance that makes cows produce more milk than normal. *page 52*

brain stem the part of the brain that controls automatic life processes. *page 28*

brain the part of the body that controls all the body's activities. *page 28*

capillaries tiny blood vessels that deliver blood to the body's cells. *page 22*

carbon dioxide a colorless, odorless gas produced when fuel is burned. *page 62*

carnivores animals that eat only meat. *page 43*

cell the smallest part of a living thing that is living. *page 18*

cell division a process that happens when one cell splits into two new cells. *page 20*

centripetal force the force that causes things to go toward the center. *page 123*

cerebellum the part of the brain that coordinates muscle activities. *page 28*

cerebrum the part of the brain in which thinking occurs. *page 28*

chemistry the study of matter. *page 90*

cholesterol a type of fat found in the human body and in foods made from animals. *page 24*

circulatory system the heart and all the blood vessels. *page 22*

compost a mixture of plant waste used to make soil richer. *page 68*

compound a substance made of two or more elements. *page 92*

conclusion the decision on whether the hypothesis is supported by evidence. *page 14*

constellation an imagined picture made of stars. *page 80*

core the center of Earth. *page 71*

coronary artery disease a disease that causes the arteries to become blocked. *page 24*

cortex the part of the cerebrum that stores information. *page 28*

crust the outer layer of Earth. *page 71*

cubic mile an area one mile long, one mile wide, and one mile high. *page 61*

DDT a chemical that kills insects. *page 51*

decibel a unit of measurement of volume. *page 117*

dissolve when one substance disappears into another to make a solution. *page 101*

electron a negatively charged particle in an atom. *page 91*

element a substance that can't be broken down into simpler substances. *page 90*

endangered species any type of animal in danger of extinction. *page 44* energy the ability to change or move things. *page 132*

environment everything that surrounds us, including air, water, soil, plants, and animals. *page 50*

erupt explode violently. *page 61*

experiment a method used to test a hypothesis. *page 13*

extinct no longer in existence. *page 42*

fault a crack in rock along which plate movement occurs. *page 72*

ferns plants with roots, stems, and leaves but no seeds. *page 47*

fiber thin threads that make up fabric and other materials. *page 68*

filament a thin, flexible wire. *page 96*

food chain a cycle in which plants are eaten by animals, which are eaten by other animals. *page 51*

fossils the remains of once-living things. *page 42*

fossil fuels coal, oil, gas, and other fuels that formed from the bodies of living things millions of years ago. *pages 50 and 134*

fruit the part of a flowering plant that contains seeds. *page 47*

galaxy a group of stars held together by gravity. *page 78*

gas the state of matter that has a volume and shape that can change. *page 94*

gene a microscopic part of a living thing that tells the living thing how to develop. *page 13*

Goldilocks Conditions the conditions needed for life as we know it. *page 78*

gravity the force that pulls any two objects together. *pages 77 and 122*

greenhouse effect the warming of Earth caused by an increase of carbon dioxide in the atmosphere. *page 62*

H_2O the chemical formula for water. *page 94*

habitat the place where an animal or plant lives. *page 43*

hairy mammoths large, fur-covered, elephant-like animals that no longer exist. *page 42*

heart attack a condition in which part of the heart dies from lack of blood. *page 24*

herbivores animals that eat only plants. *page 42*

hertz a unit of measurement of pitch. *page 117*

hypothesis a careful guess about the answer to a question. *page 13*

immune system the body's system for fighting disease-causing germs. *page 37*

immunization the process of creating resistance to particular germs. *page 38*

incinerator container where garbage is burned. *page 66*

insulation a material used to hold in heat. *page 68*

joints places where bones are linked together. *page 34*

landfill a place where garbage is buried. *page 66*

laser a device that makes a thin beam of light made up of all the same wavelength. *page 127*

ligaments tough, elastic bands that hold two bones together. *page 33*

light year the distance that light travels in one year. *page 128*

liquid state of matter that has a volume that can't change but a shape that can change. *page 94*

mantle the part of Earth between the crust and the core that is liquid and very hot. *page 71*

marrow the material inside bones that produces blood cells. *page 32*

mass the amount of material in an object. *pages 77 and 91*

matter a substance that occupies space and can be seen, sensed, or measured. *page 90*

mercury a poisonous silver-colored metal that has a low melting point. *page 96*

microscope an instrument that makes very small things look bigger. *page 18*

Milky Way the name of the galaxy where our sun and Earth are found. *page 78*

mixture a combination of two or more substances. *page 100*

molecule two or more atoms connected together. *page 94*

mosses and liverworts small plants that grow in damp places. *page 47*

nerve cell a cell with long fingerlike parts that help it send and receive messages. *page 19*

nervous system all the nerves in the body and the brain. *page 29*

neutron an uncharged particle in an atom's nucleus. *page 91*

newton a unit that measures force. *page 122*

nuclear power plant factory that changes energy from the nucleus of an atom into electrical energy. *page 133*

nuclear winter a large drop in temperature resulting from a nuclear explosion. *page 62*

nucleus an atom's center that contains protons and neutrons. *pages 91 and 132*

observe watch and read to gather information about something. *page 12*

optical fiber thin fibers of glass or plastic used to send information. *page 127*

orbit the path of a planet around the sun. *page 77*

organ part of the body, such as the heart or the skin, that does a particular job. *page 29*

organism any form of animal or plant life. *page 20*

osteoporosis a disease that causes weak bones. *page 32*

parent cell a cell that splits into two new cells. *page 20*

pH scale a measure of how strong an acid or base is. *page 107*

pitch quality of a sound. *page 117*

plate a section of Earth's crust. *page 71*

pollutant something that pollutes, especially a waste material that affects air, soil, or water. *page 106*

pollute to give off harmful substances. *page 66*

pollution the dirtying or poisoning of the environment. *page 50*

populations the numbers of a group of plants or animals that live in one place. *page 51*

prehistoric existing millions of years ago. *page 42*

prism a piece of clear glass that bends white light, showing the spectrum. *page 126*

properties qualities belonging to an individual or a thing. *page 102*

proton a positively charged particle in an atom's nucleus. *page 91*

react when chemicals change each other. *page 106*

recycling reusing something instead of just dumping it. *page 67*

reflection the effect that happens when light, sound, or heat bounces off a surface. *page 117*

reflex a quick response caused by nerves in the spinal cord. *page 29*

rotation the time it takes a planet to spin one complete turn on its axis. *page 77*

scientific methods processes for getting information and testing ideas. *page 12*

seasons yearly changes in temperature caused by the tilt of a planet's axis as it moves around the sun. *page 78*

seed cone a scaly part of a nonflowering plant that holds the plant's seeds. *page 47*

seed plants plants with roots, stems, leaves, and seeds. *page 47*

solar system our sun and its nine planets. *pages 60 and 76*

solid state of matter that has a volume and shape that can't change. *page 94*

solute the substance that disappears in another substance (the solvent). *page 101*

solution a mixture in which the substances are completely mixed and the different parts cannot be seen. *page 101*

solvent a substance, such as water, that dissolves another substance (the solute). *page 101*

sound wave a pattern of vibrations carried through air, water, and solid objects. *page 116*

specialized cell a cell that has parts that help it carry out a specific job. *page 19*

spectrograph a tool used to separate the colors of light given off by stars. *page 127*

spectrum the different colors contained in white light: red, orange, yellow, green, blue, indigo, and violet. *page 126*

spinal cord a part of the body that allows messages to travel between the brain and the body. *page 29*

state a form of matter: solid, liquid or gas. *page 94*

symbol one or two letters that stand for an element. *page 92*

tendons tough, elastic bands that attach a muscle to a bone. *page 33*

tide the rise and fall of the level of the oceans. *page 124*

toxins poisonous chemicals. *page 36*

transform change into another form. *page 133*

trillion 1,000,000,000,000. *page 128*

tsunami a large sea wave made by an underwater earthquake or volcano. *page 70*

tungsten a hard metal that has a very high melting point. *page 96*

universe space and everything in it. *page 76*

vaccines substances that make the body produce antibodies against a particular germ. *page 38*

veins blood vessels that carry blood to the heart. *page 22*

vibrate shake rapidly. *page 94*

virus a tiny part of a cell that can copy itself only by using living cells. *page 36*

volcano an opening in the earth's crust through which melted rock is forced. *page 61*

volume loudness of sound. *page 117*

water vapor the moisture in the air. *page 60*

wavelength the distance between the top of one wave and the top of the next. *page 126*

white blood cells cells that can surround and kill germs. *page 37*

yard waste clippings from lawns, raked leaves, and other plant trimmings. *page 67*

yeast a kind of fungus that is made of one cell. *page 18*

zodiac the 12 monthly constellations that follow each other across the middle of the night sky. *page 82*

Answers and Explanations

Page 3

1. **(1) cells** Choices (2), (3), and (4) are larger parts of the body than cells.

2. **(4) whole body.** In the diagram, organ systems support the whole body. Choices (1), (2), and (3) refer to smaller body parts than organ systems.

3. **(4) The body is organized from smaller to larger parts.** The first sentence gives the main idea. Choices (1), (2), and (3) are details in the paragraph but not the main idea.

Page 4

4. **(2) circulatory system.** This is stated in the last sentence. Choice (1) is incorrect because the term *blood system* is not mentioned. Choices (3) and (4) are not mentioned in the paragraph.

5. **(3) arteries through the capillaries to the veins.** The arrow on the diagram shows the blood moving from the arteries through the capillaries and to the veins. Choices (1), (2), and (4) list the blood vessels in incorrect order.

6. **(2) There are three kinds of blood vessels.** The first sentence gives the main idea. Choices (1), (3), and (4) are details.

Page 5

7. **(4) Australia** Almost all the land in Australia is covered by desert or semi-arid areas.

8. **(3) Europe** All the continents listed have deserts except Europe.

9. **(4) Africa** Choice (1) is incorrect because Africa's desert area is larger than Australia's. Choices (2) and (3) are incorrect because neither has a desert as large as Africa's.

Page 6

10. **(3) ball shaped.** The sphere is described as *round*. Choices (1), (2), and (4) are incorrect because they are not round and do not look like the sun up in the sky.

11. **(4) 99** The paragraph states that the sun makes up 99 percent of the solar system.

12. **(2) large.** Choices (1), (3), and (4) are incorrect because they do not fit the description of the sun's properties.

Page 7

13. **(3) burning coal** Choices (1), (2), and (4) are incorrect because human activities are the main cause of adding acid rain-causing chemicals to the air. The paragraph also states that volcanoes are not the main cause of acid rain.

14. **(3) 6.3 trillion ounces** The key shows which bars represent human and natural causes. The white bar shows that the amount of nitrogen oxide released by natural processes is 6.3 trillion ounces per year. Choices (1), (2), and (4) are the amounts shown by the other bars on the graph.

Page 8

15. **(1) flat mirror** Of the three diagrams, only the diagram of the flat mirror shows light reflecting but not spreading out.

16. **(1) There are three types of mirrors.**
Choices (2) and (4) are incorrect because
they are too specific to be the main idea.
Choice (3) is incorrect because the image
seems smaller than the object only in a
convex mirror.

17. **(1) her face larger than it really is** The
paragraph states that images in concave
mirrors appear larger than the objects
being reflected.

Page 9

18. **(1) solid to a liquid.** The line on the
graph shows a substance melting and then
vaporizing as temperature increases.
Choice (2) is not shown on the graph.
Choices (3) and (4) occur as temperature
decreases.

19. **(4) boiling water.** Boiling water is a gas
and is represented by the upper right part
of the graph. Choices (1), (2), and (3)
occur at lower temperatures and have less
thermal energy than boiling water.

20. **(2) thermometer.** Thermal energy (heat)
and temperature increase and decrease
together, so a thermometer can be used to
measurer thermal energy. Choices (1), (3),
and (4) do not measure temperature.

Unit 1 Lesson 1

Page 15

1. scientific method   3. experiments

2. hypothesis

4. **(1) observation.** Choices (2), (3), and
(4) are incorrect because they are different
scientific methods. Observation is looking
carefully at a problem.

5. **(2) a conclusion.** Choice (1) is incorrect
because an experiment precedes the
conclusion. Choice (3), a scientific method,
refers to the entire process, of which the
conclusion is one part. Choice (4), an
observation, occurs before, during, and
after the experiment.

6. 1. Observe. 2. State the problem as a
question. 3. Make a hypothesis.
4. Experiment. 5. Draw a conclusion.

GED Skill Strategy, *pages 16–17*

Page 16

Exercise 1: word: theory; meaning: an
explanation that is based on many experiments
and observations.

Page 17

Exercise 2:

1. Slip on a shirt. Slop on sunscreen. Slap on
a hat.

2. Sun Protection Factor—tells how strong
the sunscreen is

3. 15 or higher

Lesson 2

Page 21

1. specialized cells
2. cell division
3. microscope
4. **(4) nerve cell** Choice (1) is incorrect because blood cells look like cushions and deliver oxygen to cells. Choices (2) and (3) are incorrect because bone cells make up bones, and muscle cells make up muscles.
5. **(2) bacterium** Choices (1), (3), and (4) are all incorrect because animals, humans, and plants are many-celled organisms.
6. sugar

Lesson 3

Page 25

1. veins
2. arteries
3. capillaries
4. **(1) to the capillaries** Choice (2) is incorrect because veins carry blood back to the heart. Choice (3) is incorrect because arteries carry blood away from the heart. Choice (4) is incorrect because "used" blood goes to the liver, lungs, and kidneys for cleaning.
5. **(3) to the heart** Choice (1) is incorrect because fresh blood passes through the heart and arteries before going to the capillaries. Choice (2) is incorrect because the veins take "used" blood to the liver, lungs, and kidneys for cleaning. Choice (4) is incorrect because "used" blood goes to the liver, lungs, and kidneys for cleaning.
6. Any two: (1) Eat foods low in fat and cholesterol. (2) Eat foods high in fiber and vitamins. (3) Exercise regularly. (4) Stop smoking.

GED Skill Strategy, *pages 26–27*

Page 26

Exercise 1: Main idea: Muscles of the Upper Arm.

Page 27

Exercise 2: Details: The diagram shows that the brain has several different parts. It also shows that different parts of the brain control different body activities.

Lesson 4

Page 31

1. brain
2. brain stem
3. cerebrum
4. cerebellum
5. **(3) making sense of what you hear, see, touch, taste, and smell** Choice (1) is an activity of the digestive system. Choices (2) and (4) are activities of the circulatory system.
6. **(4) a quick response** Choice (1) is incorrect because a complicated thought does not need to result in quick action. Choice (2) is incorrect because reflexes take place in healthy people. Choice (3) is incorrect because a reflex is not a thing.
7. **(2) the bones** Choice (1) is incorrect because it is part of the circulatory system. Choice (3) is incorrect because nerves are part of the nervous system that is being protected. Choice (4) is incorrect because the eyes do not protect the nervous system.

Lesson 5

Page 35

1. marrow   2. ligament, joint   3. tendon

4. **(1) Bones are like tubes.** Choices (2), (3), and (4) are all things that would make bones heavy, not light.

5. **(2) Minerals and protein fiber are held together firmly.** Choice (1) is incorrect because marrow does not add to a bone's strength. Choices (3) and (4) are incorrect because tendons and ligaments attach bones and muscles, not make them strong.

6. There are many possible ways to answer this question. Here is an example.

   There are two types of fractures, open and closed. An open fracture is one that breaks through the skin; a closed fracture remains under the surface of the skin.

Lesson 6

Page 39

1. bacteria, viruses   2. toxins   3. Antibodies

4. **(3) to stop the growth and spread of bacteria** Choice (1) is what immunization does. Choice (2) is incorrect because the drugs are not designed to attack the symptoms. Choice (4) is incorrect because antibiotics do not work against viruses.

5. The cell bursts open.

6. The contents of the cell spill out.

GED Skill Strategy, *pages 40–41*

Page 40

Exercise 1: <u>But the skin is as important as the other organs</u>.

Exercise 2: There are three kinds of blood vessels: arteries, veins, and capillaries.

Page 41

Exercise 3: The skin has several jobs.

Lesson 7

Page 45

1. Extinct

2. Endangered

3. Carnivores

4. **(3) Tigers were once common, but now they are rare.** Choice (1) is not true. Choices (2) and (4) are details that support the main idea.

5. **(1) The jaguar is an endangered species.** Choices (2), (3), and (4) are details that support the main idea.

6. **(4) Earth's climate became colder.** Choices (1), (2), and (3) are not supported by details in the passage.

7. **(4) all of the above** Choices (1), (2), and (3) are all part of the many efforts to save endangered animals.

Lesson 8

Page 49

1. Mosses   2. Ferns   3. Seed

4. **(4) Plants have many uses.** Choices (1), (2), and (3) are incorrect because they are details that support the main idea.

5. **(3) leaves** Choice (1), roots, holds the plant in the ground and absorbs water and minerals from the soil. Choice (2), the stem, supports the upper part of the plant. Choice (4), the flower, produces seeds.

6. **(2) fruit** Choice (1), roots, holds the plant in the ground and absorbs water and minerals from the soil. Choice (3), the flower, produces seeds. Choice (4), leaves, makes food for the plant.

Lesson 9

Page 53

1. fossil fuels   2. pollution   3. DDT

4. **(1) chemical fertilizers** Choice (2) is incorrect because the crops themselves don't pollute. Choice (3), rainwater, just carries the chemicals into lakes and streams. Choice (4) has nothing to do with water pollution.

5. **(4) to increase the amount of milk** Choices (1) and (2) are incorrect because just the opposite happens when bGH is used. Choice (3) is incorrect because the milk is already safe to drink.

6. DDT first entered the food chain when it was sprayed on an insect.

7. There are many possible ways to answer this question. Here is an example.

Once DDT enters a food chain, every member of the food chain is affected.

GED Test-Taking Strategy

Page 55

1. **(4) 60 million** Choice (1) is the number of buffalo in 1900. Choice (2) is the date that there were only 1,000 buffalo left. Choice (3) is the number of square miles of prairie left in 1900.

2. **(3) A leg vein can be used to bypass a blocked artery.** Choice (1) is a detail, not the main idea. Choice (2) is incorrect because the diagram shows a repair being made to the heart, not the leg. The diagram does not show information about Choice (4).

GED Test Practice, *pages 56–58*

Page 56

1. **(3) An ecosystem can become unbalanced if a new species is released in an area.** Choices (1) and (2) are details that support the main idea. Choice (4) states the opposite of the main idea.

2. **(2) be sold as pets.** Choice (1) is incorrect and is not discussed in the paragraph. Choice (3) is incorrect. Choice (4) states the opposite of the main idea.

3. **(1) They are smaller than American turtles.** Choice (2) is incorrect. Although the paragraph does not state it directly, you can figure out from the details that the American turtles are probably stronger. Choice (3) is incorrect because the American turtles are driving the French turtles out. Choice (4) is not stated in the paragraph.

Page 57

4. **(2) Plants give off oxygen.** Choice (1) is not shown in the diagram. Choice (3) is incorrect because animals take in oxygen. Choice (4) is incorrect because the diagram does not show what plants do with the carbon dioxide they take in.

5. **(3) Plants and animals use each other's waste gases.** Choices (1) and (4) are details, not the main idea. Choice (2) is incorrect because the diagram does not show how rabbits and plants get food.

Page 58

6. **(3) You can control some muscles, but not others.** Choice (1) is incorrect because it is a detail from the paragraph. Choice (2) is incorrect because the paragraph does not include this information. Choice (4) is incorrect because only some muscles move automatically.

7. **(1) heart muscle** Choices (2), (3), and (4) are all muscles you can control. The paragraph states that the heart is controlled by muscles that work automatically.

8. **(3) when you breathe.** Choices (1), (2), and (4) are activities you think about as you do them. The paragraph states that breathing and your heart beating are automatic.

Unit 2 Lesson 10

Page 63

1. atmosphere
2. nuclear winter
3. greenhouse effect

4. **(2) helps keep Earth's temperature even.** Choice (1) is incorrect because the atmosphere protects life on Earth. Choice (3) is incorrect because sunlight does reach the surface of Earth. Choice (4) is incorrect because the atmosphere extends thousands of miles from Earth's surface.

5. **(3) carbon dioxide in the atmosphere.** Although Choices (1), (2), and (4) are gases found in the atmosphere, only carbon dioxide increases the greenhouse effect.

6. There are many ways to answer this question. Here is one example:

   Greenhouse glass allows sunlight to enter the building and then traps the heat inside. This is much like carbon dioxide in the atmosphere because it also does not allow all the heat from the sunlight to escape.

GED Skill Strategy, *pages 64–65*

Page 65

1. raining
2. Houston
3. Boston
4. Oklahoma City
5. 60s

Lesson 11

Page 69

1. Recycling   2. compost   3. landfill

4. **(2) dumping garbage in the ocean**
   Choices (1) and (3) are ways that we can reduce the amount of garbage in landfills. Choice (4) is a reason why we must look for other ways to get rid of garbage.

5. **(3) Southeast** The Southeast has 738 landfills. Choices (1), (2), and (4) are incorrect because these regions have fewer landfills than the Southeast.

Lesson 12

Page 73

1. mantle   2. fault   3. earthquake

4. **(2) Predicting where an earthquake will occur is not very difficult.** Choices (1) and (3) are incorrect because they are details. Choice (4) is not mentioned.

5. **(3) Predicting when an earthquake will occur is difficult.** Choices (1), (2), and (4) are incorrect because they are details.

6. Most of the faults are located in southern California.

7. There are many answers to this question. Here are some examples:

   San Francisco, Hollister, Bakersfield, Northridge, and Los Angeles.

GED Skill Strategy, *pages 74–75*

Page 74

Exercise 1: *Spewed* means "thrown out with force." The words around it, such as *exploded* and *violently into the air*, help explain what the word means.

Page 75

Exercise 2: There are many ways to answer this question. Here is an example.

Curbside recycling is separating regular garbage from items that can be recycled, which are collected separately from the regular garbage.

Exercise 3: There are many ways to answer this question. Here is an example.

A *radiosonde* is a tool used to gather weather information.

Lesson 13

Page 79

1. Big Bang   2. orbit   3. Milky Way

4. **(3) sun** The first paragraph states that more than 99 percent of the solar system's mass is in the sun. Choices (1), (2), and (4) are incorrect.

5. **(4) 365 days.** The paragraph states that an orbit is a planet's year. One year is about 356 days. Choices (1), (2), and (3) are all shorter than a year.

6. **(2) Venus.** The picture on page 77 shows that Venus takes 243 Earth days to rotate one time. This is longer than the planets listed in Choices (1), (3), and (4).

Lesson 14

Page 83

1. constellations

2. zodiac

3. Southern Cross

4. **(1) close together.** The paragraph states that gravity pulls gas and dust close together. It also states that the star gives off heat and light. Choices (2) and (3) do not relate to these two pieces of information. Choice (4) is the opposite of what is true.

5. **(3) January** The other choices are not supported by the map.

GED Test-Taking Strategy

Page 85

1. **(3) Tornado season is different in different areas.** Choices (1), (2), and (4) are details that support the main idea.

2. **(1) crust, mantle, outer core, inner core** Choices (2) and (3) list the layers out of order. Choice (4) lists the layers from inside to outside.

GED Test Practice, *pages 86–88*

Page 86

1. **(1) Earth is made up of moving plates.** Choices (2), (3), and (4) are details that support the main idea.

2. **(3) melted.** The paragraph states that Earth's plates float on molten rock. Choices (1), (2), and (4) do not describe something that plates can float on.

3. **(3) where one plate rides up over another plate.** Choice (1) describes most places in the world. Choices (2) and (4) refer to other processes.

Page 87

4. **(4) which places are likely to be damaged by earthquakes.** Choice (1) is incorrect because the map does not show this. Choices (2) and (3) may seem to be true based on the map, but they are not the main idea.

5. **(3) Dallas** Choices (1), (2), and (4) are incorrect because these cities are more likely to have earthquake damage than Dallas.

Page 88

6. **(1) A hot spot is like a blowtorch.** Choices (2) and (4) are details from the paragraph. Choice (3) is incorrect because the paragraph does include this information.

7. **(3) a hot area inside Earth.** Choices (1) and (2) are things that hot spots are being compared to. Choice (4) is incorrect because it is not what the paragraph is about.

Unit 3 Lesson 15

Page 93

1. element
2. compound
3. atomic
4. protons, electrons

5. **(3) two or more elements.** Choice (1) is incorrect because a compound contains two or more different elements. Choice (2) is incorrect because elements are all different. Choice (4) is incorrect because atomic numbers don't make up compounds.

6. **(2) protons and neutrons.** Choices (1), (3), and (4) are incorrect because electrons are located outside the nucleus.

7. **(3) a compound.** Choices (1) and (2) are incorrect because water is made up of two elements. Choice (4) is incorrect because protons alone don't make up a compound.

Lesson 16

Page 97

1. tungsten
2. state
3. liquid

4. **(3) move back and forth.** Choice (1) is incorrect because the paragraph says that the increase in vibration moves molecules out of place. Choice (2) is incorrect because nothing in the paragraph mentions breaking down. Choice (4) is incorrect because the molecules don't change as the ice cube melts.

5. **(4) a tiny wire that lights up.** Choices (1) and (2) are incorrect because a filament is not a kind of metal. It is made of metal. Choice (3) is incorrect because

there is nothing in this paragraph about liquid in a light bulb.

6. **(3) lined up in an orderly way** Choices (1), (2), and (4) are incorrect because the paragraph explains that the molecules of ice, water vapor, and liquid water are all the same.

7. **(4) all of the above** As stated in the paragraph, the states of matter are solid, liquid, and gas.

GED Skill Strategy, *pages 98–99*

Page 98

Exercise 1:

1. carbon dioxide
2. makes
3. heat

Page 99

Exercise 2:

1. Adding salt lowers the freezing point of water.
2. The water stays in a liquid state.
3. Salt does not mix easily with the water.

Lesson 17

Page 103

1. mixture
2. solution
3. dissolves

4. **(3) a mixture.** Choice (1) is incorrect because you cannot see or easily separate the parts of a solution. Choice (2) is incorrect because air is a solution. Choice (4) is incorrect because a solute is part of a solution.

5. **(1) Sugar is the solute, and water is the solvent.** Choices (2) and (3) are incorrect because water is the solvent. Choice (4) is incorrect because sugar is the solute.

6. **(2) mixing it with silver** Choice (1) is incorrect because simply melting gold makes it softer than it is as a solid. Choices (3) and (4) are incorrect because the lesson does not discuss mixing gold with these materials.

GED Skill Strategy, *pages 104–105*

Page 104

Exercise 1:

1. Name, Chemical Formula, Where Found
2. $C_6H_8O_7$ 3. Hydrochloric acid

Page 105

Exercise 2:

1. Number of animals, scale; Kind of animal
2. lake trout   3. 20 mayflies

Page 109

1. rain   2. base   3. acid
4. **(2) pollutants in the air react with rain water.** Choices (1), (3), and (4) are not causes of acid rain.
5. **(3) the killing of plants and animals.** Choices (1), (2), and (4) are not effects of acid rain.
6. **(3) North America and Europe.** The map shows no acid rain damage in South America, Africa, and Asia, so choices (1), (2), and (4) are incorrect.

7. **(4) both (1) and (3).** Choices (1) and (3) are both supported by the information. The paragraph tells you that traffic and factories contribute to acid rain. The map shows that much damage by acid rain has occurred in Europe. Neither the map nor the paragraph support choice (2).

Page 111

1. **(4) inside a factory smokestack** Choices (1), (2), and (3) are not discussed in the paragraph. The paragraph states that scrubbers clean factory smoke.
2. **(3) water** Choices (1) and (4) are incorrect because gas and smoke are being cleaned. Choice (2) is incorrect because the scrubber helps keep the air clean.

GED Test Practice, *pages 112–114*

Page 112

1. **(2) oxygen** Choices (1) and (4) are not discussed in the paragraph. Choice (3) is incorrect because iron oxide forms as a result of rusting.
2. **(1) wearing away.** Choice (2) is incorrect because corrosion is only one kind of chemical reaction. Choices (3) and (4) mention things that are the opposite of corrosion.
3. **(3) a rubber bicycle tire** Choices (1), (2), and (4) are all made of metal that can rust when exposed to air.

Page 113

4. **(4) bringing the substance to its melting point.** Choice (1) is incorrect because the amount of heat needed to melt differs from substance to substance. Choices (2) and (3) are incorrect because not all substances melt under these circumstances.

5. **(3) sucrose** Choices (1), (2), and (4) all have lower melting points and thus lower freezing points.

6. **(2) oxygen** Choices (1), (3), and (4) all have a higher melting point than oxygen does.

Page 114

7. **(4) the sugar in grain** Choices (1), (2), and (3) are each a result of fermentation. Choice (4) is a cause of fermentation.

8. **(4) carbon dioxide.** Choice (1) is incorrect because sugar is broken down to form the bubbles. Choice (2) is incorrect because alcohol is not a gas. Choice (3) is incorrect because yeast produces the bubbles.

Unit 4 Lesson 19

Page 119

1. decibel   2. Reflection

3. **(4) both (1) and (2)** Choice (3) is incorrect because the sound waves themselves are invisible.

4. **(1) Sound waves reflect off walls and other surfaces.** Choices (2) and (3) don't explain the reason for an echo. Choice (4), therefore, is incorrect.

5. **(2) energy level.** Choices (1) and (3) are incorrect because pitch and frequency are measured in hertz, not decibels. Choice (4) is incorrect because *intensity* is a way to describe a sound wave, not another word for *sound wave.*

6. **(4) send.** Choice (1) is incorrect because the eardrum does not stop the sound waves. Choice (2) is incorrect because the eardrum doesn't make any new waves. Choice (3) is incorrect because if the waves were absorbed, they couldn't continue on to the inner ear.

GED Skill Strategy, *pages 120–121*

Page 120

Exercise 1:

1. the speed of sound, temperature

2. The speed of sound is greater at 20 degrees Celsius than at 0 degrees Celsius. The graph shows that the speed of sound increases with temperature.

Page 121

Exercise 2: 1,000 meters; If the runner maintains the same speed, the line will continue in the same way.

Exercise 3: The gas's volume will be greater at 500 kelvins. The graph shows that a gas's volume increases with temperature.

Lesson 20

Page 125

1. gravity, centripetal force
2. newtons
3. **(3) 80 N** Choices (1), (2), and (4) are not supported by the graph.
4. **(4) The force is about 675 N more at Earth's surface.** Since that graph shows that the force is 720 N at the surface, about 675 N more than the 45 N at 19,200 km, choices (1), (2), and (3) are incorrect.
5. **(2) The moon would move out into space.** Earth's gravity keeps the moon circling around it. Choices (1), (3), and (4) are incorrect because they contradict Choice (2).

Lesson 21

Page 129

1. prism
2. trillion
3. spectrum
4. wavelength
5. **(4) both (2) and (3)** Choice (1) is incorrect because a prism alone can't cause a rainbow. Choices (2) and (3) are incomplete answers.
6. **(2) Light waves travel faster than sound waves.** Choice (1) is not true. Choices (3) and (4) are not the cause of seeing lightning before hearing thunder.
7. **(4) both (1) and (3)** Choices (1) and (3) are incomplete answers by themselves. Choice (2) is incorrect because the light waves appear as separate colors.

GED Skill Strategy, *pages 130–131*

Page 130

Exercise 1: How do we hear? <u>The outer ear receives sound waves and directs them toward the eardrum. The eardrum vibrates and passes the sound waves to the inner ear. The inner ear contains a liquid that is moved by the sound waves. Hairlike structures in the inner ear feel the moving liquid and send messages to the brain.</u> These messages are what we call hearing.

Page 131

Exercise 2: (2) They are all about how hearing occurs. Choice (3) is incorrect because only the facts about the inner ear mention a liquid. Choice (1) is incorrect because only the last two sentences are about the brain.

Exercise 3: (2) collect sounds so you can hear. Choice (1) is incorrect because the ear doesn't make sounds. Choice (3) is incorrect because the ear does not block sound.

Exercise 4:

1. **(2) Objects would fly off into space.** Choices (1) and (3) are incorrect because the passage mentions nothing about the speed at which objects would move.
2. A. The paragraph states that we fall down when we trip.

   B. The last sentence states that things such as people, buildings, trees, and the oceans do not fly off into space.

Lesson 22

Page 135

1. nucleus
2. fossil fuels
3. electrical energy
4. chemical energy
5. **(4) D** Choices (1), (2), and (3) are incorrect because these points on the graph represent less heat energy than point D does.
6. **(1) 5 seconds** Choices (2), (3), and (4) are incorrect. To get the answer, count the number of squares along the Time axis between 0 degrees Celsius (4 seconds) and 300 degrees Celsius (9 seconds).

GED Skill Strategy, *pages 136–137*

Page 136

Exercise 1:

1. You will not see a rainbow. Although there is both sunlight and raindrops, the sun is not behind you so a rainbow won't form.
2. You will not see a rainbow. Clouds on a cloudy day do not allow enough sunlight to pass through to form a rainbow.

Page 137

Exercise 2: There are many possible ways to answer the question. Here is an example.

I agree with the prediction. There would be many more dust particles in the air, so the blue light would be scattered more quickly, and we would see mainly red and orange light, like at sunset.

GED Test-Taking Strategy

Page 139

1. **(2) Greater volumes of water take more time to boil.** Choice (1) is not addressed by the graph. Choices (3) and (4) are only details on the graph.
2. **(1) Laser light is in phase.** Choices (2), (3), and (4) contain facts that are not true according to the paragraph.

GED Test Practice, *pages 140–142*

Page 140

1. **(1) how lasers at checkout lines work** Choices (2), (3), and (4) are details listed in the paragraph.
2. **(2) The store's computer would not read the price code.** Choice (1) is correct only if the computer can read the code. Choices (3) and (4) are incorrect because different equipment would not be able to read a missing price code either.
3. **(2) lasers make the checkout process faster.** The laser reads the price quickly. This can shorten the time and increase the accuracy of checking out. Choices (1), (3), and (4) are not supported by the paragraph.

Page 141

4. **(4) An object has the same mass on Earth as on the moon.** Choice (1) is incorrect because the graph shows different scales for mass and weight. Choice (2) is incorrect because the graph shows smaller numbers (values) for masses

than for weights. Choice (3) is incorrect because the graph shows that an object has a different weight on the moon than on Earth.

5. **(3) 6.5 newtons on Earth.** Choice (1) gives the mass of the object on the moon. Choices (2) and (4) are incorrect because they correspond to about 0.1 newton and 1.5 newtons on the moon, respectively.

6. **(4) about 12 newtons** Choice (1) is incorrect because 1.2 newtons correspond to less than 0.2 kilogram. Choice (2) is incorrect because 2 newtons correspond to about 0.2 kilogram. Choice (3) is incorrect because 8 newtons correspond to about 0.8 kilogram.

Page 142

7. **(1) Power companies get energy from different sources.** Choices (2), (3), and (4) are not stated in the paragraph.

8. **(2) nuclear energy** Choices (1) and (3) are incorrect because fuel and water are used to make electricity. Choice (4) does not involve splitting atoms.

Science Posttest, *pages 143–150*

Page 143

1. **(3) in the center of the tree trunk** The paragraph states that the older rings are on the inside of a stump. Choices (1) and (2) are not discussed. Choice (4) is incorrect because the younger rings are found here.

2. **(2) 5 years.** The diagram shows five pairs of light and dark bands. Choices (1), (3), and (4) are incorrect because they would relate to 1, 10, and 20 pairs of bands, respectively.

3. **(3) Every year trees produce a double layer of wood.** Choices (1), (2), and (4) are details that support the main idea.

Page 144

4. **(4) summer** Long-day plants begin to flower when the days are long, as in summer. Choices (1), (2), and (3) refer to times when the days are short.

5. **(2) during the long days of summer.** Choices (1) and (3) are incorrect because the diagram shows the geranium does not bloom during short days. Choice (4) is incorrect because the geranium blooms only when grown during the long days.

6. **(4) The amount of daylight affects how plants grow.** Choices (1), (2), and (3) are details in the paragraph.

Page 145

7. **(4) Dallas** Of the cities listed, only Dallas is covered by the slanted lines that show rain on the map.

8. **(2) in the 20s** If you look at Chicago on the map, you will see that it is in a clear band area that has temperatures in the 20s.

Page 146

9. (1) rainy, in the 50s San Francisco is shown covered by the slanted lines that mean rain. It is also in a shaded area that has temperatures in the 50s.

10. (2) shadow. According to the first paragraph, an eclipse has to do with shadows. Choices (1) and (4) relate to shape, which is not discussed in the paragraph. Choice (3) is not discussed.

11. (1) the moon The moon is not visible from Earth during a lunar eclipse. Choices (2), (3), and (4) can be seen during a lunar eclipse.

12. (1) the sun. During a solar eclipse, the sun is not visible from Earth.

Page 147

13. (1) their properties Each element has different characteristics, or properties. Choices (2), (3), and (4) are not discussed.

14. (2) oxygen The tallest bar on the bar graph is labeled *oxygen*. Choices (1), (3), and (4) refer to bars that are smaller than the one labeled *oxygen*.

Page 148

15. (2) shiny. The definition of *luster* is given in the second sentence, after the dash (—). Choices (1), (3), and (4) are incorrect because *luster* means "shiny."

16. (3) electrical wires Choices (1), (2), and (4) are all items that would not be good conductors of heat or electricity.

Page 149

17. (1) oil The line graph shows that oil would reduce friction most. Choices (2) and (3) have medium and high friction. Choice (4) is not shown.

18. (3) medium The left side of the line graph shows that the amount of friction using water as lubrication would be medium.

19. (2) makes them last longer. Choices (1), (3), and (4) are incorrect because oiling a machine has the opposite effect.

Page 150

20. (4) atmosphere This unit of measure is shown along the bottom axis of the graph. Choice (1) refers to the space the gas occupies, not pressure. Choice (2) is not a unit of measure. Choice (3) is the unit of measure for volume.

21. (2) 1 atmosphere You must find the point on the curved line that shows 1 liter. Then you can find the number of atmospheres along the bottom axis that would intersect with this point.

22. (2) The gas takes up less space. The line in the graph slopes down as the pressure increases. The volume does not increase so Choice (1) is incorrect. The volume changes, so Choice (3) is incorrect. Choice (4) is incorrect because you cannot tell the exact volume of the gas based on the information given in the question.